HOW TO
TROUBLESHOOT &
REPAIR ANY SMALL
GAS ENGINE

PAUL DEMPSEY

TAB BOOKS Inc.
Blue Ridge Summit, PA 17214

29720

FIRST EDITION

THIRD PRINTING

Printed in the United States of America

Reproduction or publication of the content in any manner, without express permission of the publisher, is prohibited. No liability is assumed with respect to the use of the information herein.

Library of Congress Cataloging in Publication Data

Dempsey, Paul.
    How to troubleshoot and repair any small gas engine.

    Includes index.
    1. Internal combustion engines, Spark ignition—
Maintenance and repair. I. Title.
TJ790.D44   1985   621.43′4   85-14748
ISBN 0-8306-0967-9
ISBN 0-8306-1967-4 (pbk.)

HOW TO
TROUBLESHOOT &
REPAIR ANY SMALL
GAS ENGINE

Other TAB Books by the Author

No.   817   *How To Repair Diesel Engines*
No.   968   *How to Convert Your Car, Van, or Pickup to Diesel*
No. 1687   *How to Repair Briggs & Stratton Engines—2nd Edition*

Contents

Introduction   **vii**

1 Engine Basics   **1**

Nomenclature—Operation—Dimensions and Measurements—Heat Lubrication—Maintenance

2 Ignition   **23**

Timing—Simple Alignment—Point Gap Adjustment—Point-
To-Cam Adjustment—Insufficient Output—Conventional
Systems—Solid-State Systems

3 Carburetors and Fuel Systems   **67**

Operation—Types—Adjustments—Filters—Governors—
Fuel Pumps—Mechanical Pumps

4 Rewind Starters   **117**

Side Pull—Service Procedures—Briggs & Stratton—Vertical
Pull

5 Electrical System   **155**

Starting Circuits—Charging Systems

6 Engine Mechanics   **188**

Diagnosis—Scope of Work—Cylinder Head—Valves—
Pistons and Rings—Cylinder Bores—Connecting Rods—
Assembly Crankshafts—Camshafts—Main Bearings—
Seals—Governor Mechanisms—Oiling Systems

Index   **259**

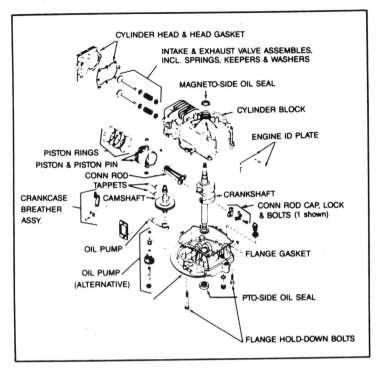

Fig. 1-2. Internals of a four-cycle, single-cylinder, vertical crankshaft engine of the type that is used to power rotary lawnmowers.

from the combustion chamber into the crankcase.

● Piston pin—ties the piston to the connecting rod; also known as the wrist pin.

● Connecting rod—connects the piston and the explosive force generated against the piston crown to the crankshaft. Usually aluminum, although two-cycle engines sometimes employ a forged steel connecting rod.

● Crankshaft—acts in concert with the connected rod to convert reciprocating piston movement into rotary motion suitable for turning wheels or driving other machines. Often cast iron, sometimes cast or forged steel.

● Main bearings—two on a single cylinder engine to support the crankshaft at both ends. May be in the form of sleeve, or plain, bearings or antifriction (ball, roller needle) bearings.

● Cylinder block—major casting, includes cylinder bore and crankcase cavity. This cavity holds the oil supply for four-cycle engines and forms part of the two-cycle induction tract, channeling air, fuel, and oil from the carburetor into the combustion chamber. Either aluminum or cast iron.

• Flange—secondary casting that closes off one side of the crankcase cavity. Depending upon configuration, may be known as side cover or oil pan.

• Valves—salient characteristic of four-cycle engines, usually mounted in cylinder block (as shown) in which case the engine is described as a side-valve or L-head design. Overhead valve (ohv) with valves mounted in head, above piston crown, are also manufactured.

—Intake valve: opens to admit air-fuel mixture from carburetor to combustion chamber.

—Exhaust valve: opens to permit egress of exhaust gases from combustion chamber to atmosphere.

• Camshaft—acts through tappets, or valve lifters, to cam the valves open at appropriate places in the operating cycle. Driven at half engine speed by the crankshaft. Valves are closed by springs.

• Oil pump—fitted on many four-cycle engines to transfer oil under pressure to the bearings.

Figure 1-3 shows a typical two-cycle block assembly for comparison with the four-cycle in Fig. 1-2. Note the absence of many parts, including intake and exhaust valves, camshaft and oil pump. Valve functions are accomplished by the piston that opens and closes ports in the cylinder bore. Many two-cycle engines, including the one illustrated, also employ a reed, or leaf, valve between the carburetor and crankcase. Two-cycle lubrication is by means of small quantities of oil mixed with the fuel.

Another difference between the block assemblies shown is that the four-cycle unit employs a vertical crankshaft, mounted "north-south" in the block. This configuration is used in applications such as rotary lawnmowers, where the driven member is below the block assembly. The Clinton two-cycle uses a horizontal crankshaft, typical of most industrial engines. Other than a change in carburetor-fuel tank position and some modification to four-cycle crankcases for oil storage, there is no significant mechanical difference between crankshaft configurations.

OPERATION

Internal combustion engines operate in a cycle of four events

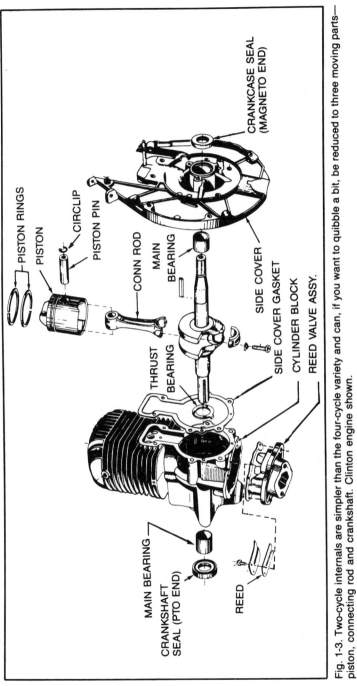

PISTON RINGS

PISTON

CIRCLIP

PISTON PIN

CONN ROD

MAIN BEARING

THRUST BEARING

SIDE COVER

SIDE COVER GASKET

CYLINDER BLOCK

REED VALVE ASSY.

CRANKCASE SEAL (MAGNETO END)

MAIN BEARING

CRANKSHAFT SEAL (PTO END)

REED

Fig. 1-3. Two-cycle internals are simpler than the four-cycle variety and can, if you want to quibble a bit, be reduced to three moving parts—piston, connecting rod and crankshaft. Clinton engine shown.

that take place in the area above the piston. These events are *intake* of fuel and air, *compression* of the charge, ignition and subsequent *expansion* and *exhaust* of spent gases. Four-cycle engines require four upward or downward strokes of the piston to complete the full cycle. Two-cycle engines compress events into two strokes of the piston, or one crankshaft revolution.

Four Cycle

Figure 1-4 shows the sequence of piston and valve movement during the four events. The piston moves downward during the intake stroke, evacuating the cylinder above it. Air and fuel enter around the open intake valve to fill this void. The exhaust valve is closed. At the lower limit of piston travel, called bottom dead center (bdc), the intake valve closes and the piston begins to move upward in the compression stroke. The piston rounds top dead center, the spark plug fires, and the piston descends on the power (ex-

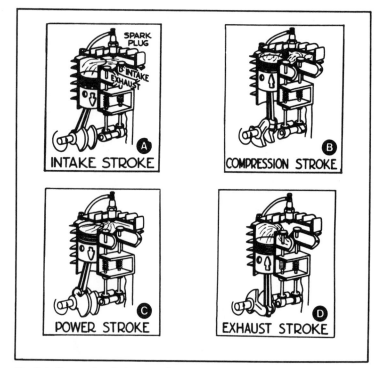

Fig. 1-4. Conventional piston engines operate in a cycle, or sequence, of four events—intake, compression, power and exhaust. In a four-stroke-cycle engine, this sequence requires two complete crankshaft rotations or four piston strokes.

pansion) stroke. Both valves remain closed to contain the force of the explosion. Past tdc again, the piston moves upward, forcing the spent gases out through the open exhaust valve. The intake valve will remain closed until the piston reaches tdc of the exhaust stroke and a new cycle begins.

So much for theory. In practice, events are not so neatly apportioned, one event to each piston stroke. Gases have interia (that is, they resist acceleration) and, once set in motion, are reluctant to stop moving. Both valves open early—well before their respective strokes—and remain closed late. Ignition occurs early, some 10 to 15 crankshaft degrees before top dead center, so that full combustion pressure has time to develop before the piston retreats on the downstroke.

Theory and practice do conform at one place in the cycle. At top dead center on the compression stroke, both valves are closed, although a fraction of a degree of crank movement in either direction will crack one or the other valve. This bit of information is useful when "blind timing" an engine, as explained in Chapter 6.

Two Cycle

The central element of a two-cycle engine is the piston, which functions as a double-acting compressor and, in conjunction with ports in the cylinder bore, as a shuttle valve. The mixture is compressed both in the combustion chamber above the piston, and in the crankcase below it. Piston movement uncovers one or more transfer ports, connecting the crankcase with the cylinder bore, and, somewhat later in the cycle, exhaust ports to vent the spent gases to the atmosphere. In addition, some engines use the piston to open the intake port from the carburetor to the crankcase.

If all this seems a bit complicated, Fig. 1-5 will help to clarify matters. Keep your eye on the pea. The engine shown is typical of most industrial and light-utility types in that it uses a deflector piston and a reed, or leaf, valve.

In Fig. 1-5 (top) the piston is descending on the down stroke. Exhaust gases from the previous combustion are flowing over the piston crown through the exhaust port on the left of the drawing. At the same time, the port labeled "intake" (most mechanics would call it a transfer port) is open, connecting the area above the piston with the crankcase. The crankcase and the air-fuel-oil mixture in it are subject to slight pressurization as the piston falls and fills the block cavity. This pressure—on the order of 7 psi—is sufficient

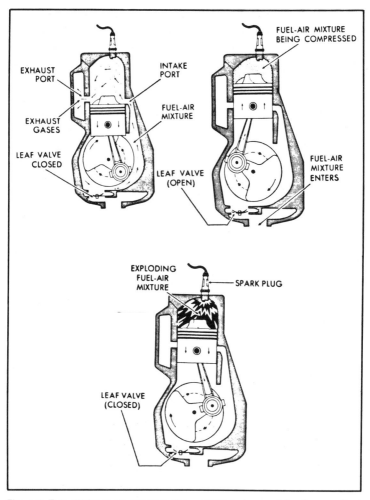

Fig. 1-5. Two-stroke-cycle engines also operate in terms of the four cardinal events (intake, compression, power, and exhaust), but action is speeded up to take place in a single crankshaft revolution or two piston strokes. (Courtesy OMC).

to force the fuel charge through the open "intake" port and into the cylinder bore.

In Fig. 1-5 (bottom) the piston has progressed past bdc and its upward movement compresses the mixture in the cylinder bore. Both "intake" and exhaust ports are closed. The same piston movement that pressurizes the bore depressurizes the crankcase. This partial vacuum causes the reed valve to open, admitting a fresh air-fuel charge to the crankcase. As the piston approaches tdc, the spark plug fires, igniting the mixture. The piston is driven downward to uncover the exhaust port and then the transfer port.

The piston shown in this example has a peaked crown, acting as a miniature "Continental Divide," to separate the gas streams. The intake charge passes through the transfer port, strikes the steep side of the piston deflector, and rebounds upward—driving the exhaust gases out before it. This system, known as a cross-flow scavenging, is relatively inexpensive to manufacture and is used in many small engines. Its chief disadvantages are poor scavenging (i.e., much of the exhaust remains in the cylinder, especially at low rpm). loss of fuel through the exhaust, and excessive piston weight.

Loop scavenging was developed some years ago in Germany and has since appeared in a number of industrial engines (Fig. 1-6). This system employs intake ports arranged radially around much of the bore diameter and angled upward to direct the incoming charge at the roof of the combustion chamber. The air-fuel streams rebound off the roof, meet and combine in a miniature tornado to

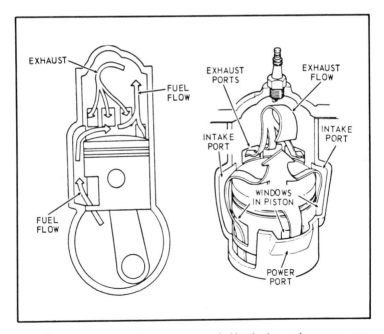

Fig. 1-6. A loop-scavenged engine uses angled intake (or, perhaps more correctly, transfer) ports imparts a swirl to the intake charge that scrubs exhaust gases from the chamber. (Courtesy OMC).

sweep the cylinder clean of exhaust products. The whirling mass of air and fuel particles has kinetic integrity and very little of it bleeds out the exhaust port.

Loop scavenging requires extremely precise foundry work, because slight errors in port location or entry angle will destroy the effect. This might help explain why some seemingly identical two-cycle engines run more smoothly and more powerfully than others.

Another area that has caught the attention of designers is the valve between the crankcase and carburetor. The reed valve illustrated functions automatically in response to the pressure differential between the crankcase and carburetor. It may improve low speed torque output because there is no possibility of charge reversal. Some small European engines use a third port, located at the base of the cylinder bore and in communication with the crankcase. The piston uncovers this port near the top of the stroke and closes it on the downstroke (before the transfer port opens). This system is simple and absolutely reliable; however, part of the crankcase charge squirts back into the carburetor as the valve closes. This can cause a ragged idle and give rise to a fog of fuel and oil at the carburetor intake.

Several imported motorcycle engines and Tecumseh water-cooled outboards employ one or another form of rotary valve. The valve is keyed to the crankshaft and has part of its face outaway. As the piston approaches tdc, the cutaway aligns with the intake port to admit a fresh charge into the crankcase.

DIMENSIONS AND MEASUREMENTS

The bore, or the diameter of the cylinder, and the stroke, or the distance the piston travels between dead centers, are the basic engine dimensions. Together, bore and stroke give the displacement, or swept volume of the engine. The formula is

$$\text{Displacement} = \frac{d^2 \times 8sn \times 3.14}{4}$$

d = diameter of bore
s = length of stroke
n = number of cylinders

The Kohler two-cylinder model K582 has a 3.50-inch bore and a 3.00-inch stroke. Plugging these values into the equation:

$$\text{Displacement} = \frac{3.50 \times 3.50 \times 8 \times 3 \times 2 \times 3.14}{4}$$

$$= 57.7 \text{ cubic inch}$$

Because bore and stroke dimensions were given in inches, displacement is expressed in cubic inches. If the engine were dimensioned metrically, displacement would be in cubic centimeters (cc). Multiplying the cc value by 0.061 gives displacement in cubic inches. To work the conversion the other way, multiply cu in. by 16.39.

Compression Ratio

Compression ratio (cr) is the ratio between the volume of the cylinder at bottom dead center compared with the volume remaining above the piston at top dead center. Within limits set by the threshold of detonation and the requirement for reasonable tractability, the higher the cr, the more power the engine develops.

Detonation occurs in most side-valve engines when compression ratio is raised beyond 7.5 or 8l to 1. Because of a more favorable combustion chamber shape, some overhead valve engines tolerate a higher cr, but rarely more than 8.5 to 1 with modern, low-octane gasoline. The phenomenon is quite complex, but basically detonation consists of an abrupt rise in combustion pressure *after* normal ignition has commenced. During combustion, the flame front moves outward from the electrode gap at the spark plug, progressively igniting the air-fuel mixture. Detonation occurs when part of the mixture, compressed and heated by the expanding flame front, spontaneously ignites, generating terrific force that vibrates the connecting rod like a tuning fork to produce an audible ping. Severe detonation will puncture a piston in a few minutes and moderate, transitory detonation will shorten engine life.

In addition to keeping the compression ratio conservative, there are various techniques to keep detonation in check. One way is to increase the speed of flame propagation so that normal combustion can occur before the tag ends of the mixture absorb enough

11

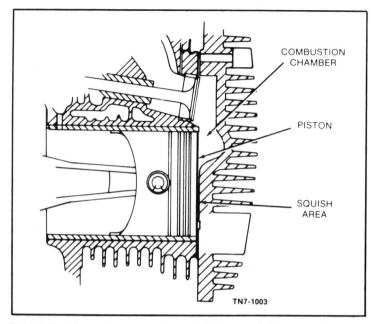

Fig. 1-7. Squish area increases mixture turbulence and allows some increase in compression ratio. (Courtesy Onan).

heat to explode on their own. This can be done by increasing the turbulence of the air-fuel mixture with a squish area above the piston (Fig. 1-7). As the piston approaches top dead center, that part of the mixture above it is compressed and "squishes" into the larger chamber.

From a mechanic's point of view, detonation can be controlled by avoiding lean mixtures (which burn slowly), impressing on operators that the engine should not be lugged (wide-open throttle at low rpm reduces turbulence) and by doing everything possible to keep temperatures within design limits. Cooling fins should be periodically cleaned, ignition timing should be checked and carbon buildup in the combustion chamber should be removed during tuneups. Two-cycle fuel should be mixed exactly as the manufacturer recommends. Whatever an extra dollop of oil might do to prolong cylinder bore life, it also reduces the fuel's octane rating and increases the likelihood of detonation.

Engine modifications should be approached cautiously. Industrial utility engines tolerate a few thousandths skimmed off the cylinder head to re-establish the gasket seal, but serious milling to raise the compression ratio is not acceptable.

Performance Data

Horsepower is a term with a certain ambiguity attached to it. The concept combines two distinct quantities: 1 hp is the ability to lift 550 pounds. 1 foot in 1 second, or 33,000 pounds, 1 foot in 1 minute. Notions of work—which in the technical sense means to exert force through distance—and time are combined to produce a measure of the rate at which work is accomplished.

There are several varieties of horsepower. *Taxable hp* is a legal term, created during the early days of automobiles to provide authorities with a yardstick for taxation. Thus, the 2CV Renault, a mini-car favored by a segment of the French population, develops two taxable hp (chevaux vapor). Output by the more common definition is 18 or so horsepower.

Indicated hp (ihp) is an engineering term, arrived at by calculation and meaning the power produced across the piston face. It does not include frictional and other parasitic losses.

The most common measure—and the one usually quoted in factory literature—is *brake horsepower* (bhp), determined by measuring crankshaft output on an engine brake, or dynamometer. But few things are a simple as they first appear, and a dynamometer test is no exception. External variables, such as ambient air pressure, temperature, and humidity affect engine power—as does accessory load. Most U.S. manufacturers conform to Society of Automotive Engineers (SAE) standard J607, which calls for a sea-level test at 60° F with standard accessories in place. Procedures minimize the possibility of dyno "loading" and high flash readings. European manufacturers test under DIN standards, which are more conservative than SAE. Power outputs can be adjusted to the American practice by multiplying the European PS (pferd stark) rating by 0.986.

Engines, especially those that are sold internationally, sometimes carry a wattage rating, which is the electrical equivalent of bhp. One bhp equals 745.7 watts or 0.7457 kilowatts (Kw).

Most industrial plants are warranted to develop at least 85% of full rated bhp upon shipment and 95% of rating after run-in. Carburetor and timing adjustments might be needed to achieve these figures and loads developed by nonstandard accessories must be deducted. In addition, power output decreases about 3.5% for each 1000 feet of altitude above sea level and 1% for each 10° F rise in temperature above 60° F.

Torque is related to horsepower, but is a distinct concept. It is a measure of instantaneous twisting force on the crankshaft, which in this country is usually expressed as pounds of force exerted on the end of a lever 1 foot long. A 33,000-pound force on a 1-foot lever for 1 minute equals 1 horsepower:

$$\text{horsepower} = \frac{6.38 \times \text{torque} \times \textit{rpm}}{33,000} = \frac{\text{torque} \times \textit{rpm}}{5,252}$$

From a theoretical point of view, torque is a function of cylinder displacement and brake mean effective pressure (bmep). The latter is the average combustion pressure during the power stroke.

$$\text{torque} = \frac{\textit{bmep} \text{ (psi)} \times \text{displacement (cu. in.)}}{150}$$

Horsepower is the index of an engine's ability to pull a load through time. In motor vehicles this means miles per hour, in pumps, gallons per minute, and in generators kilowatt-hours. Torque, on the other hand, translates as tenacity, as the ability to keep slogging as sudden loads shift the rpm curve into the high torque area.

Bmep acting on piston area produces torque. Bmep depends upon compression ratio and volumetric efficiency, or the mass of air-fuel mixture in the cylinder (Fig. 1-8). Most engines develop best volumetric efficiency and, consequently, peak torque at low to moderate speeds.

Torque appears in the horsepower equation as the force component; rpm is the time component. Horsepower curves start small because there is little rpm and torque has not yet peaked. Horsepower builds with speed and continues to rise, riding on rpm, after torque has reached its maximum value. But rpm can only carry it so far because internal friction increases with rpm, following a rough approximation of the square law. At some point near full-governed speed, horsepower peaks and begins to decline.

Everyone wants higher performance, and especially if it can be extracted without a weight or engine size penalty. But there are other tradeoffs as exhibited by torque, horsepower and fuel consumption curves for Kohler 241 and K341 engines (Fig. 1-9). The K241 is a conservatively rated workhorse with a 3.25-inch bore and

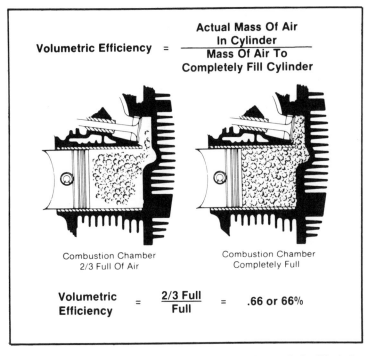

Volumetric Efficiency $=$ $\dfrac{\text{Actual Mass Of Air In Cylinder}}{\text{Mass Of Air To Completely Fill Cylinder}}$

Combustion Chamber
2/3 Full Of Air

Combustion Chamber
Completely Full

$$\text{Volumetric Efficiency} = \frac{2/3 \text{ Full}}{\text{Full}} = .66 \text{ or } 66\%$$

Fig. 1-8. Volumetric efficiency is the index of how well the cylinder fills during the intake stroke. Complete filling, that is, with the charge pressurized at 1 atmosphere, equals 100 percent volumetric efficiency. (Courtesy Onan).

2.88-inch stroke for a displacement of 23.85 cubic inches. Compression ratio is a moderate 6.2 to 1. The K341 is the performance version of the same engine, bored to 3.75 in. and strokes to 3.25 in. The engine has a 35.90 cubic inch displacement and a compression cr of 7.3 to 1, which stands near the upper limit for a side-valve industrial engine. However it does perform, developing 16 bhp and slightly more than 38 foot pounds of torque. The K241 produces only 10 hp and 21.5 foot pounds.

As you might expect in a free-lunchless world, the larger, more powerful engine burns more fuel. But the shape of the torque curve tells a more revealing story. The K341 uses bigger valves than the K241, raises them further off their seats, and keeps them open longer. These changes improve volumetric efficiency and torque production, but at some cost in flexibility. Torque peaks at 2600 rpm—300 rpm more than K241 peak—and falls sharply as speed increases. In contrast, the K4 K241 curve remains essentially flat through the rpm range, dropping by less than 3 footpound at 3600

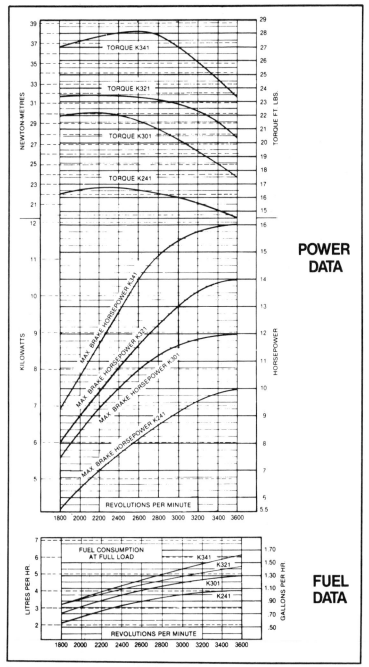

Fig. 1-9. Performance data for three Kohler single-cylinder engines.

16

rpm. As industrial engines go, the K341 is a "peaky" engine, developing good, but rpm-sensitive torque.

HEAT

Large quantities of heat are developed in the combustion chamber where normal operating temperatures can reach 2000° F. Unfortunately, only about a third of this heat is available for useful work at the crankshaft; about a third goes out the exhaust and, another third heats the engine, either from exposure to combustion gases or through friction between moving parts.

Because pure aluminum melts at about 1400° F and the lubricating properties of motor oil become problematic at about 400° F, the engine needs to be cooled in some way. The irony is that, by cooling, a thermal gradient is set up between the combustion chamber and the surrounding metal which invites more heat to be rejected during the next combustion event. Hence, the interest in high-melting-point, ceramic-engine parts and dry lubricants.

Contemporary small engines use three mechanisms to maintain internal temperatures at tolerable levels. The primary mechanism is a flywheel-mounted fan that blows or, more rarely, pulls cooling air over the cylinder head and cylinder barrel. Four-cycles depend heavily on the lube oil in the sump to cool the underside of the piston and lower end bearings; two-cycles use the incoming air-fuel-oil mixture to the same effect. An excessively gasoline-rich air-fuel mixture can also be used to quench combustion chamber temperatures at high speed or under heavy load.

Air-cooling virtues include utter simplicity, rapid engine warm-up, and virtual freedom from routine maintenance. But the freedom is not absolute and the mechanic should periodically dismantle the shrouding and clean the fins. The hub screen, fitted to most flywheel fans, might also require attention. In addition, everything possible should be done to control combustion chamber temperatures. The combustion chamber should be routinely decarbonized, timing should be advanced no more than the manufacturer specifies, and a slightly rich carburetor power setting never hurts. Of course, all shrouding—including the insignificant-appearing square of tin that covers the downstream cylinder barrel on most engines—must be in place before the unit is started.

LUBRICATION

Motor oil has multiple functions. It reduces friction by inter-

17

posing a fluid film, that can be no more than a few molecules thick, between moving parts. Heavier films, measuring in the thousandths of an inch, help to cushion forces acting on the crankshaft bearings and adhere to the cylinder walls, forming a gas-resistant seal with the piston. Local hot spots are quenched with oil that dissipates the heat load into the relatively cool crankcase. It is said that some engines depend upon oil to transfer as much as 40% of the total thermal load. Another function of oil is to disperse solids that will then be trapped in a filter. Oil also contains additives that resist ash formation (particularily important in two cycles where most of the oil is burnt in combustion), corrosion, and foaming.

Ignition problems cause most complaints, but the real engine killer is dirty oil. Two cycles have an advantage in this regard: the oil supply and lubricating film will be skimpy, but it is at least clean. Four-cycle engines recirculate their oil from a constantly diminishing sump. It is up to the operator to change the oil at recommended intervals (usually 25 operating hours or 50 hours if the engine has an oil filter) and to more or less continually check the oil level. Once the level drops, the remaining oil overheats and carburizes into an abrasive slurry.

Use the type and grade of motor oil recommended by the manufacturer for the engine in question. Synthetic and multiviscosity oils are usually not acceptable, with some manufacturers going out of their way to warn against 10W-40. Two-cycles generally require 30-weight oil, mixed in a separate container (to assure complete dispersal of the lubricant in the fuel) to exact proportion indicated on the engine instruction plate. Straight-viscosity 30-weight is the almost universal recommendation for four-cycles operated in above-freezing weather. Some manufacturers reluctantly allow a thinner, multiweight oil in lower temperatures, but others insist upon such hard-to-obtain grades as 5 weight. API classification SF, the highest grade currently available, is preferred. Don't combine cheaper oils with longer-than-recommended change intervals.

Engine assembly work is crucial from a lubrication point of view. Parts should be antiseptically clean and assembled sopping wet with 30-weight motor oil. Additives are not recommended either by the factories or by most mechanics. In spite of claims to the contrary, "Miracle-in-a-can" products do not exist.

Lubrication systems that distribute and collect oil in four-cycle engines take three patterns. The simplest is the splash system that flings about the crankcase by means of a scoop or rotating slinger.

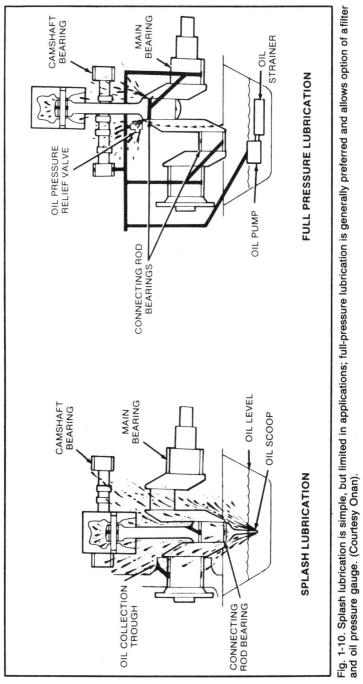

Fig. 1-10. Splash lubrication is simple, but limited in applications; full-pressure lubrication is generally preferred and allows option of a filter and oil pressure gauge. (Courtesy Onan).

19

Full-pressure systems employ a pump to provide positive lubrication to every journal bearing, while semi-pressure systems use a pump to supply remote areas and splash for the rest. In all cases, oil returns to the sump by gravity. See Fig. 1-10.

MAINTENANCE

Small, air-cooled engines are anachronistic devices, making about the same demands on their owners as automobiles did 50 years ago. Engine life, measured as operating hours between overhauls, is relatively short. Several years ago, a U.S. manufacturer of inexpensive two-cycle engines suggested that 500 hours was a reasonable figure. Engines ran that long in the lab. A maintenance schedule published by a well-known industrial engine maker indicates that after 1000 hours all bets are off.

Table 1-10. Maintenance Schedule.

Operating hours	Service
4	Check and top off crankcase oil level
8	Service air cleaner (Chapter 3)
25	Change oil (engines without oil filters)
50	Change oil and filter Clean and regap spark plug (Chapter 2) Check battery electrolyte level (Chapter 5) Clean battery terminals (Chapter 5) Replace paper filter element (Chapter 3) Torque mounting bolts Lubricate controls and control cables
100	Tune-Up time: Clean cylinder head and cylinder barrel fins, removing shrouding for access Service ignition (Chapter 2) Rebuild carburetor (Chapter 3) Adjust governor and control linkages (Chapter 3) Clean two-cycle exhaust ports (Chapter 6)
200	Clean crankcase breather (Chapter 6) Decarbonize combustion chamber (Chapter 6) Inspect valve seats and valve lash (Chapter 6)
1000	Disassemble for overhaul or rebuild

To put these numbers into context, the heavy equipment industry generally considers one hour of stationary operation to be the equivalent of 20 miles in a motor vehicle. There are documented cases of oil field pumping units—single-cylinder, gas-fired, air-cooled engines—operating continuously for 40 years with only minor maintenance. Mercedes-Benz industrial diesels, somewhat detuned and running at about half rated speed, typically clock 5000 hours without more than an injector change.

Even though schedule maintenance cannot extend design life, it can delay the inevitable and help eliminate surprises during the interim. Most important are four-cycle lubrication systems and air filters for both engine types. Oil levels should be topped off daily and changed at recommended intervals or more often under adverse operating conditions. Air filters should be cleaned and re-oiled at least once every duty day. Air filter gaskets should be replaced at the first sign of wear and possible leakage. Polyurethane filters require special attention. Before each startup, the filter element should be removed from its housing and gently knealed to redistributed oil throughout the foam. Paper filter elements should be replaced at 50-hour intervals, and more frequently in dusty environments.

A good maintenance schedule includes the cumulative items shown in Table 1-1. Repeat the 25-hour service at 50 hours, the 50-hour service at 100 hours, and so on.

Chapter 2

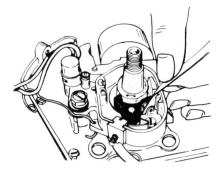

Ignition

Ignition system difficulties are of two kinds: Improper timing, and Insufficient spark.

TIMING

The spark plug should fire at some preordained point before the piston reaches top dead center (tdc) on the compression stroke. The amount of spark advance varies with engine type and intended use, but generally ranges from 15 to 30° of crankshaft rotation before tdc.

Small utility engines have, for the most part, fixed timing. Engines in industrial or leisure-product roles, where power and fuel economy is important, cannot rely on factory tolerances and always have provision for in-field timing adjustments.

SIMPLE ALIGNMENT

A number of American-made engines employ Phelon or Wico magnetos with elongated slots at the stator-mounting flange. These slots allow the whole assembly to be moved a few crankshaft degrees relative to the piston. Tecumseh provides punchmarks on the stator flange and engine block (Fig. 2-1). When these marks are aligned, the engine is considered to be in time. Jacobsen uses travel limits established by the elongated mounting holes as timing references. From the magneto side of the engine, loosen the two

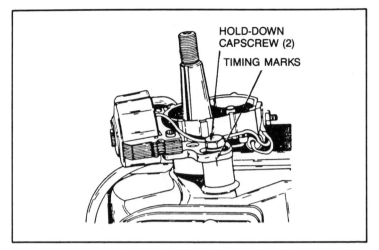

Fig. 2-1. Tecumseh engines are timed by aligning marks on the magneto stator and stator pedestal. If the stator is replaced—as when substituting a magneto from another engine—all bets are off, and the engine must be timed as shown in Fig. 2-5.

magneto hold-down capscrews and rotate the magneto—as far as slots will allow—counterclockwise for 20-inch. Snow Jet engines. Retighten the capscrews. The drill is similar for all other Jacobsen products except that the magneto is turned fully clockwise, establishing an ignition advance of 20° at crankshaft, equivalent to 1/8 inch of piston travel before tdc.

West Bend, which uses a Wico magneto 5 on slotted mounts, varies magneto position from full left to full right, depending upon model and engine rotation. Original timing should be referenced before disassembly.

Timing procedures as just discussed are somewhat removed from the real problem (which is to synchronize ignition with piston movement). Ignition in conventional systems occurs when the points break open and are no longer conductive. Most manufacturers coordinate point break with a stationary timing mark. A few manufacturers go further and coordinate point break with a measured piston travel distance before top dead center.

The four timing procedures described in the remainder of this chapter detail how this is done. The first two subsections—"Point Gap Adjustment" and "Point-To-Cam Adjustment"—describe how contact point geometry is varied to synchronize ignition with a fixed timing mark. The "Timing Light" subsection describes how both point and solid-state ignitions are timed dynamically (while the

gap specification. For example, if manufacturer calls for a 0.020-inch gap, the actual adjustment can range between 0.017 and 0.023 of an inch.

POINT-TO-CAM ADJUSTMENT

There is a limit to how much point gap variation the ignition system can tolerate. One way around this constraint is to arrange

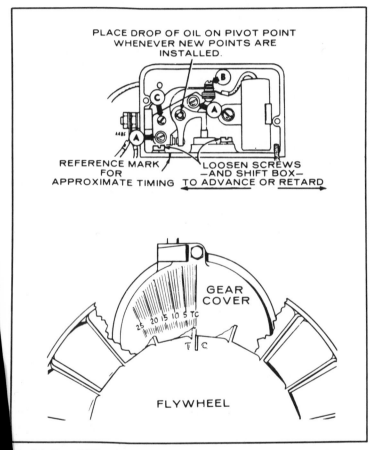

Fig. 2-2. Onan CCK and CCKA series engines are timed by moving the breaker box relative to the camshaft. Top dead center is indicated by the letters TC, standard industry practice. But because Onan timing specs vary with engine model and optional equipment, no specific timing mark is provided. CCK engines are timed 19° before top dead center (btdc). When a strobe light is used, CCKA electric-start engines with auto advance are timed at 24 btdc; electric start CCKA engines without the advance mechanism are retarded slightly 20 btdc.

engine is running). The fourth subsection "Point Break/Piston Position" shows how ignition is synchronized with piston position for many magneto-equipped engines.

POINT GAP ADJUSTMENT

Some familiarity with ignition contact points and their terminology is required to understand the following paragraphs. Readers are referred to the "Contact Points" section of this chapter.

Point gap has considerable influence on ignition timing, particularly when points are driven at half engine speed by the camshaft. Making the gap wider allows the points to remain open longer and advances timing relative to piston position. By the same logic, narrowing the gap reduces point duration and retards timing. This relationship holds for all point-equipped engines, but not all manufacturers specify it as part of their timing drill. Those that do have generally arranged matters so that the points are mounted outside of the flywheel with timing marks on flywheel rim and engine crankcase.

A test lamp (or meter) is needed to accurately determine when points open. The lamp should have its own power source when use on magneto engines. On engines with conventional battery and c ignitions, the engine battery can be used for the lamp. The gene procedure is as follows:

- Set ignition point gap to manufacturer's specs.
- Disconnect primary wire to magneto coil, because th is grounded and would short the test lamp.
- Connect one test lamp lead to engine ground and th to the moveable, or "hot," point arm.
- Locate timing marks, that might which may require of flywheel shroud (Wisconsin Robin) or inspection plu Kohler models).
- Bring piston up on compression stroke so that tim approximately align.
- Turn the crankshaft about half a revolution a direction of normal rotation.
- Slowly turn the crankshaft in normal rotation u indicates contact points have separated.
- Adjust point gap to synchronize point separatio mark alignment. Range of permissible adjustment v manufacturer, but is usually not more than ± 15%

for the contact points to move as an assembly relative to their actuating cam. Wico and Phelon under-flywheel magnetos (described earlier) have this feature, because their slotted mounts allow rotation relative to the cam. But for the examples described, the timing marks were used as final arbiters, and no attempt was made to determine when the points opened. The discussion here concerns those systems that combine point assembly movement with determination of point opening.

Figure 2-2 shows how ignition timing is varied on Onan CCK series engines. Point gap is adjusted to the specified 0.020 of an inch, a light is connected across the points, flywheel timing marks are aligned, and the point box is moved left or right to initiate point break.

Timing Light

Any engine with external timing marks—that is, marks that are visible when the engine is assembled and running—can be timed with a strobe light. Engines with solid-state ignition systems or with automatic advance mechanisms *must* be timed with a strobe light. Solid-state trigger circuits cannot be opened to determine the moment of firing and, therefore, must be timed while the engine is running. Nor can automatic advance mechanisms, which may be mechanical or electronic, be definitively bench tested.

An inexpensive neon timing light is all that is required. Xenon lights, which are favored by automobile mechanics, require an external power source that might not be available on a magneto-equipped engine.

Gap the points (where applicable) to factory specs and connect the strobe to the spark-plug wire. Twin cylinder engines are normally set up to fire alternately, 90° apart for two-strokes, 180° apart for four-strokes. When this is the case, one cylinder, known at No. 1, is the timing referent and the strobe must be connected to its spark-plug lead. A few twins fire simultaneously, either for bona fide geometric reasons or because one cylinder receives a phantom spark during its exhaust stroke. The strobe can be connected to either lead.

Engines with fixed advance are timed at idle speed, although speed can be increased to detect point bounce and possible misfires. Illuminated by a strobe, timing marks appear stationary and it is simple to bring them into alignment by rotating the magneto or solid-state module relative to the crankshaft.

The drill for automatic advance mechanisms is harder to

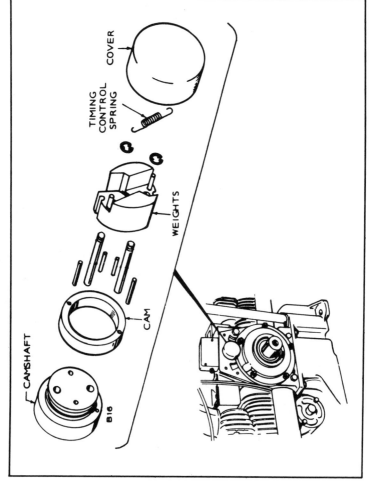

Fig. 2-3. Onan automatic advance mechanism may occasionally need cleaning or, at even longer intervals, spring replacement. Engines so equipped should be timed while running 1500 rpm minimum.

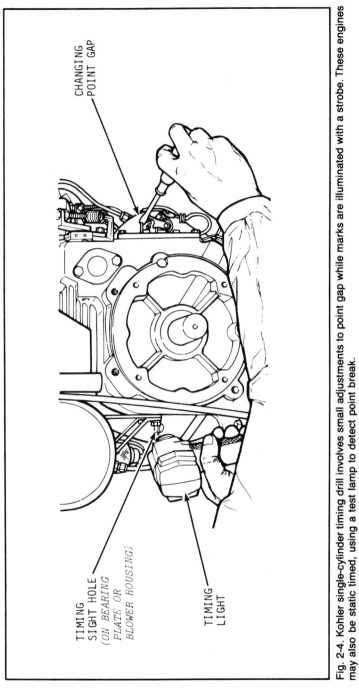

CHANGING
POINT GAP

TIMING
SIGHT HOLE
*(ON BEARING
PLATE OR
BLOWER HOUSING)*

TIMING
LIGHT

Fig. 2-4. Kohler single-cylinder timing drill involves small adjustments to point gap while marks are illuminated with a strobe. These engines may also be static timed, using a test lamp to detect point break.

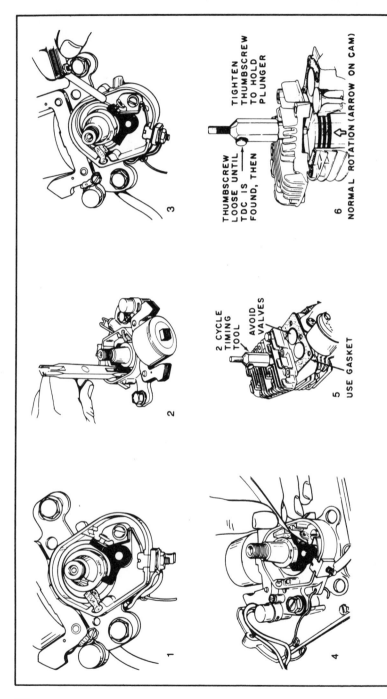

3

THUMBSCREW
LOOSE UNTIL
TDC IS
FOUND, THEN

TIGHTEN
THUMBSCREW
TO HOLD
PLUNGER

6

NORMAL ROTATION (ARROW ON CAM)

2

2 CYCLE
TIMING
TOOL

AVOID
VALVES

5

USE GASKET

1

4

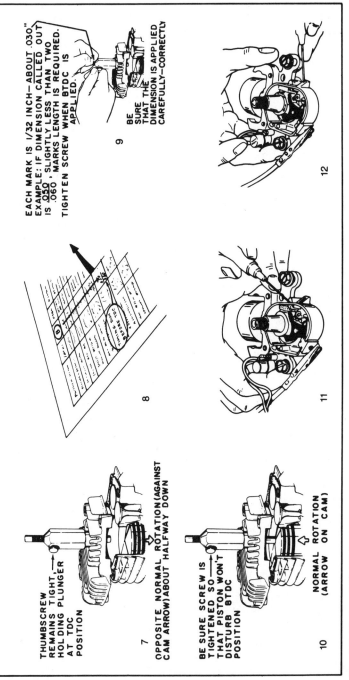

7 THUMBSCREW REMAINS TIGHT, HOLDING PLUNGER AT TDC POSITION

OPPOSITE NORMAL ROTATION (AGAINST CAM ARROW) ABOUT HALFWAY DOWN

8

9 EACH MARK IS 1/32 INCH—ABOUT .030". EXAMPLE: IF DIMENSION CALLED OUT IS .050, SLIGHTLY LESS THAN TWO .060 MARKS LENGTH IS REQUIRED. TIGHTEN SCREW WHEN BTDC IS APPLIED.

BE SURE THAT THE DIMENSION IS APPLIED CAREFULLY—CORRECTLY

10 BE SURE SCREW IS TIGHTENED SO THAT PISTON WON'T DISTURB BTDC POSITION

NORMAL ROTATION (ARROW ON CAM)

11

12

Fig. 2-5. Timing drill details.

generalize. Some manufacturers (e.g., Sachs) provide two timing marks, one for idle rpm and the other for full advance. Others provide only an idle mark, trusting the mechanics to keep rpm below the point of automatic advance. A few, such as Onan, provide a timing specification at full automatic advance. This means that the engines in question should turn at least 1500 rpm. See Figs. 2-3 and 2-4.

Point Break/Piston Position

Timing methods thus far discussed assume the accuracy of at least one timing mark. This is a significant assumption, because timing marks are at the far end of a series of toleranced parts that begins with the piston crown. Consequently, high-performance engines, that operate in a narrow window between loss of power from too much retard and meltdown from too much advance, are timed by synchronizing point separation with piston position. While any point-equipped engine (that specifications are available for) can be timed in this manner, the practice is more prevalent with two-stroke engines.

Figure 2-5 illustrates the procedure. A timing tool (available through Tecumseh dealers) is shown, although a dial indicator mounted vertically over the spark plug port can give even greater accuracy.

Timing drill for Tecumseh products can be applied to other point-equipped engines when piston travel before tdc specifications is available.

1. Install new point set and condenser.
2. Align points for full contact.
3. Rotate crankshaft to cam points full open and adjust gap to specification.
4. Burnish point contact surfaces with cardboard to help assure easy startup.
5. Install two-cycle timing tool.

In Fig. 2-5, the tool is used on a four-cycle engine where the spark-plug port is offset from the cylinder bore. Consequently, it is necessary to remove head bolts and position the head and timing tool over the piston dome. Note that the original (compressed) gasket remains in place. A new head gasket will be used for assembly.

6. Slowly turn the crankshaft in the direction of normal rotation while observing tool plunger. The plunger will rise as tdc

is approached, appear almost stationary at tdc and fall as the piston descends. When apparent tdc is found, lock the plunger with the thumbscrew and cycle the piston through again. Slight resistance should be felt as the piston top just grazes the plunger tip. Bring the piston to tdc on compression stroke.

7. Reverse crank rotation for about 90°.

8. Find timing specification for engine in question.

9. Apply specification to the timing tool by extending plunger to a specified length past tdc. Tighten the thumbscrew securely.

10. Slowly turn the crank in the normal direction to bring the piston top into light contact with the plunger tip.

11. Loosen the magneto flange capscrews and install the timing light.

12. Rotate the stator until the points crack open. With the light still connected, tighten the stator cap screws and verify that the engine timing remains as set. Disconnect the light, remove the timing tool, pin prick the new timing marks on the stator and block, and assemble remaining engine parts.

INSUFFICIENT OUTPUT

Ignition system troubleshooting should be prefaced by spark plug replacement because the spark plug is the weakest component in the system. If this does not clear up the problem, proceed to test voltage output and spark consistency.

Voltage can be measured with an automotive oscilloscope, but mechanics working on small engines almost universally use an air-gap test. Figure 2-6 shows the typical procedure. The high tension lead is held so that its terminal is about 1/8 inch from the spark-plug terminal. The engine is cranked and the quality of spark is observed. A fat, blue spark that cracks audibly indicates that all is well with the system. White, spindly sparks might not have voltage to start the engine and reddish sparks usually mean burnt points and, possibly, condenser failure.

There is an exception. Magnetron solid-state ignitions, standard on Briggs & Stratton engines since August 1982, do not produce an impressive spark at cranking speed. To test voltage output, remove the spark plug, hold the high-tension lead near a block or cylinder-head ground and crank vigorously. In general, any spark means that the system is okay.

In most instances, an ignition system that delivers healthy cranking voltage continues to operate normally at higher speeds. If you want to test for misfiring, connect a strobe light across the

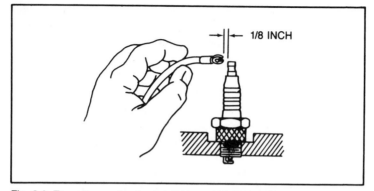

Fig. 2-6. To verify spark output, hold bared high-tension terminal about 1/8 of an inch from the spark plug center terminal or cylinder head and crank the engine. Spark should be thick and blue.

high tension lead and run the engine up in stages to governed top speed. Observe light flicker checking for slips in the pattern.

Warning: Looking directly at Xenon lamps can cause retina damage. Reflect the light off a polished surface.

CONVENTIONAL SYSTEMS

Two conventional ignition systems—battery and coil, and magneto—are encountered. Because both systems were developed in parallel, both employ similar components and, with exception of some troubleshooting procedures, can be discussed together.

Battery & Coil

Found on a number of industrial engines and motorcycles, battery and coil systems have three great advantages. Voltage output is highest during starting when it is needed most, individual components can be isolated and replaced, and primary current is present whenever the key is on (which simplifies troubleshooting).

Theory. Figure 2-7 is a schematic of a battery and coil system that, in this case, delivers voltage to two spark plugs. The primary, or battery-voltage, circuit connects the battery with ignition coil, breaker points, and condenser. The secondary, or spark-voltage, circuit connects coil and spark plugs.

When the ignition switch is *on* and contact points are closed, the battery discharges into the primary circuit, through heavy primary windings in the coil and through the stationary contact point to ground. When points open, primary current is no longer grounded

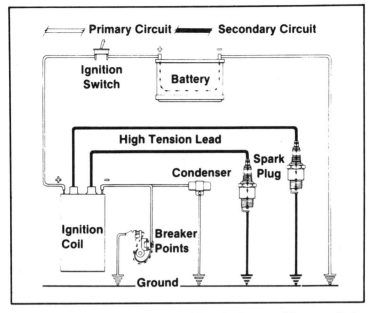

Fig. 2-7. Battery and coil circuitry for a Kohler simultaneous firing two-cylinder engine. If the cylinders fired out of phase, two coils would be used (typical Japanese motorcycle practice) or a distributor would be interposed between the high-tension lead and the spark plugs. Single-cylinder system are wired as shown with, of course, one high- tension lead and spark plug deleted.

and ceases to flow. An inductive reaction in the coil causes a high voltage spike to be generated in the secondary lead. This voltage finds ground through the spark plugs and engine block. The condenser, or capacitor, helps prevent arcing across the points.

Troubleshooting. Begin by checking the battery state of charge. Specific gravity should be 1.280 at 80° with no more than 0.05 difference between highest and lowest cell reading (Fig. 2-8). It is also good practice to monitor cranking voltage of electric-start engines (Fig. 2-9). The battery should produce at least 75% of its rated voltage during several seconds of starter drain.

Check the point gap. Wear on the rubbing block (the bearing surface between the movable point arm and the ignition cam) eventually absorbs the gap, which is typically specified at 0.020 of an inch for American small engines (0.015 of an inch on West Bend) and 0.012-0.016 of an inch for foreign makes.

Next, check primary circuit continuity with a test lamp rated at battery voltage or, barring that, a screwdriver. With the ignition switch on, turn the flywheel until the points open. Connect the lamp

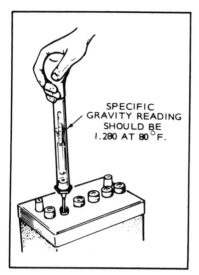

SPECIFIC
GRAVITY READING
SHOULD BE
1.280 AT 80°F.

Fig. 2-8. The first step in battery and coil troubleshooting is to determine the state of the charge of the battery with a hydrometer (Courtesy Onan).

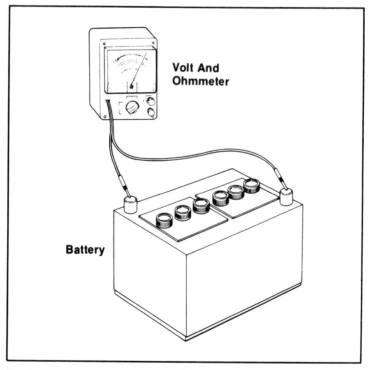

Volt And
Ohmmeter

Battery

Fig. 2-9. Battery capacity—as distinguished from state of charge—is also important for electric-start engines. The starter motor takes precedence and, if the battery is too small, will rob the ignition system of voltage.

between the moveable ("hot") point arm and ground. The light should come on. If it does not come on, trace the circuit back to its source, successively connecting the lamp to the negative coil terminal, positive terminal, ignition switch and battery terminals. Eventually the circuit break will be found.

If, on the other hand, the lamp lights when connected across the hot point arm and ground, turn the flywheel until the points close. The lamp should go out because since the hot point is now grounded through the stationary point. If it continues to burn, point contacts are oxidized and should be cleaned or replaced.

If both of these tests are negative, that is, if primary voltage is present at the moveable point arm with the points open and absent when points are closed, the problem is best solved by substitution. In order of frequency, the failed component is:

- Contact points.
- Condenser (replace together with contact points).
- Coil.
- High tension lead.

Magneto Systems

A magneto can be thought of as a battery and coil system with a high-voltage, permanent-magnet generator replacing the battery. The system is compact, self-contained, and provides a spark intensity roughly proportional to rpm. Before the advent of electronics, racing engines were universally fitted with magnetos.

Theory. Figure 2-10 shows the parts arrangement of a typical magneto, although another configuration—with points and condenser mounted remotely from the coil—is also encountered. The primary circuit consists of primary windings in the coil, breaker points, condenser, and may include a kill switch that shuts down the engine by shorting primary voltage to ground. The secondary circuit consists of secondary coil windings, high tension lead, and spark plug.

Figures 2-11 and 2-12 illustrate magneto operation about as well as drawings can. As the flywheel turns, a magnet sweeps the coil, energizing its primary windings. Some 300V is produced this way and flows through the closed points and engine ground to the coil, as shown in Fig. 2-11. In the next drawing, the points open. Primary voltage collapses together with the magnetic field associated with that voltage. Secondary windings are energized to some 12,000V and the spark plug fires.

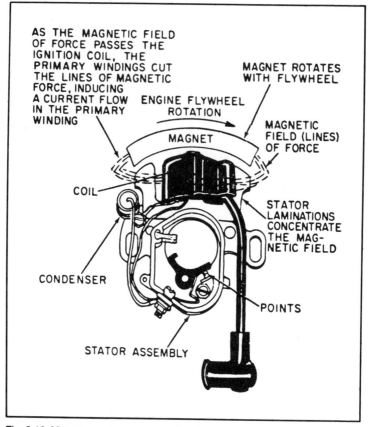

AS THE MAGNETIC FIELD OF FORCE PASSES THE IGNITION COIL, THE PRIMARY WINDINGS CUT THE LINES OF MAGNETIC FORCE, INDUCING A CURRENT FLOW IN THE PRIMARY WINDING

MAGNET ROTATES WITH FLYWHEEL

ENGINE FLYWHEEL ROTATION

MAGNETIC FIELD (LINES) OF FORCE

MAGNET

COIL

STATOR LAMINATIONS CONCENTRATE THE MAG-NETIC FIELD

CONDENSER

POINTS

STATOR ASSEMBLY

Fig. 2-10. Magneto parts arrangement and nomenclature. The cam that opens the contract points is omitted for clarity.

Troubleshooting. A magneto is heavily dependent upon parts geometry, and particularly the position of the magnets relative to the coil at the moment of point break, known as *edge distance*. For troubleshooting purposes it is sufficient to verify that the flywheel or magnet rotor key is true and that its keyway is neither wallowed or cracked. It should also be noted that rotary lawnmower engines may not start if the blade is missing or loose, because the lack of blade inertia affects edge distance.

Magnetos develop high primary currents and, consequently, point life is short. This is particularly true if the system operates off the crankshaft to generate a spark each revolution. Because the condenser affects point life and spark intensity, change both parts at the first sign of trouble.

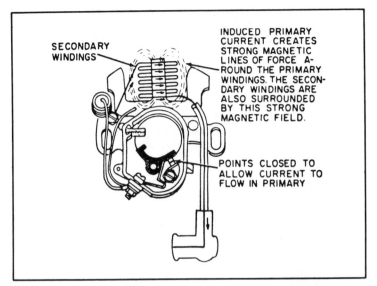

SECONDARY WINDINGS

INDUCED PRIMARY CURRENT CREATES STRONG MAGNETIC LINES OF FORCE A-ROUND THE PRIMARY WINDINGS. THE SECONDARY WINDINGS ARE ALSO SURROUNDED BY THIS STRONG MAGNETIC FIELD.

POINTS CLOSED TO ALLOW CURRENT TO FLOW IN PRIMARY

Fig. 2-11. Points are closed, completing the primary circuit which is energized by the interaction of magnetic force with primary coil windings.

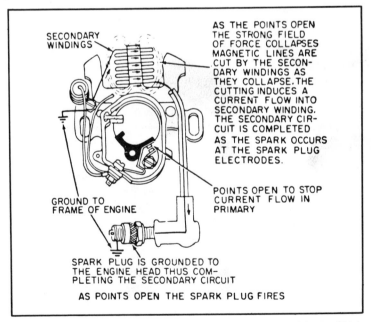

SECONDARY WINDINGS

AS THE POINTS OPEN THE STRONG FIELD OF FORCE COLLAPSES MAGNETIC LINES ARE CUT BY THE SECONDARY WINDINGS AS THEY COLLAPSE. THE CUTTING INDUCES A CURRENT FLOW INTO SECONDARY WINDING. THE SECONDARY CIRCUIT IS COMPLETED AS THE SPARK OCCURS AT THE SPARK PLUG ELECTRODES.

POINTS OPEN TO STOP CURRENT FLOW IN PRIMARY

GROUND TO FRAME OF ENGINE

SPARK PLUG IS GROUNDED TO THE ENGINE HEAD THUS COMPLETING THE SECONDARY CIRCUIT

AS POINTS OPEN THE SPARK PLUG FIRES

Fig. 2-12. Points open, breaking the primary circuit continuity. A burst of high voltage is induced in the secondary and goes to ground through the spark plug gap.

Check wiring insulation for possible damage. Pay special attention to the kill switch wire that can be fouled by carburetor controls. While it is impossible to tally all the things that can go wrong with various magnetos, one special case deserves mention. Small European motorbikes are often wired with the stop lamp in series with the ignition coil. Should the lamp fail, the coil is denied ground and the whole system shuts down.

The coil is the last component to be suspected because it is the most reliable and most expensive to replace. Coils can be checked with the appropriate equipment, but the surest check is to substitute a known good unit.

Flywheel Removal and Installation

Small engine magnetos, almost by definition, use the flywheel as the magnet carrier. Most mount contact points, condenser and coil under the flywheel, but some employ remotely mounted points driven at half engine speed by the camshaft. Battery and coil systems often incorporate under-flywheel coils for charging purposes. In any event, the mechanic should become thoroughly familiar with flywheel removal, inspection, and installation.

Small engine flywheels are secured by a crankshaft nut, a tapered fit, and a key. Most engines you will encounter use a standard right-hand thread (overhand and left to loosen). If there is any doubt about the matter, check the manufacturer's manual. Briggs & Stratton engines are a special case because units with rewind or impulse starters secure the flywheel with a light metal starter clutch assembly. The company provides a wrench (part No. 19114) and a socket (part No. 19161), either of which will allow the starter clutch assembly to be removed and installed without damage (Fig. 2-13). Flywheel fasteners on other engines can be removed with the appropriate metric or American socket.

As also shown in Fig. 2-13, Briggs & Stratton provides a flywheel holder, suitable for 6 3/4-inch OD and smaller wheels. However, its utility is limited on vertical shaft engines—the holder itself must be somehow secured—and most mechanics prefer a strap wrench (Fig. 2-14). It is bad practice to secure the assembly by blocking one of the flywheel air vanes, because the vane can snap off or, worse, can crack and fly off when the engine is running.

A lockwasher is usually present under the crankshaft nut. Note that for Briggs & Stratton the concave (dished inward) side of the lockwasher is assembled next to the flywheel.

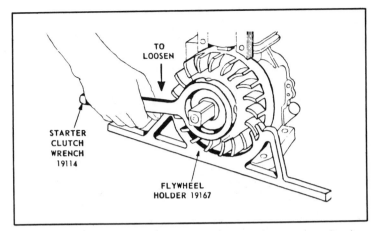

Fig. 2-13. A Briggs & Stratton starter clutch wrench is almost a necessity when servicing these engines. The company also provides a socket wrench for the same purpose.

There are at least four ways to separate flywheel and crankshaft.

● Heavy, iron wheels can usually be jimmied off with two screwdrivers inserted between the back side of the wheel and the crankcase.

● Most aluminum flywheels require a *hub puller*, which takes purchase in bolt holes provided in the flywheel hub (Fig. 2-15). Do not use a jaw-like gear puller that hooks over the flywheel rim.

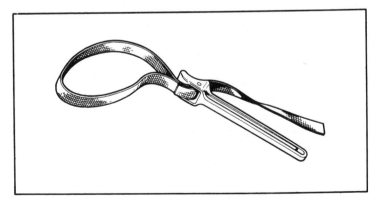

Fig. 2-14. A strap wrench, such as the one shown here and available from Tecumseh as part No. 670305, is the preferred tool for holding the flywheel. Rotary lawnmower wheels can be secured by blocking the blade with a short two by four.

41

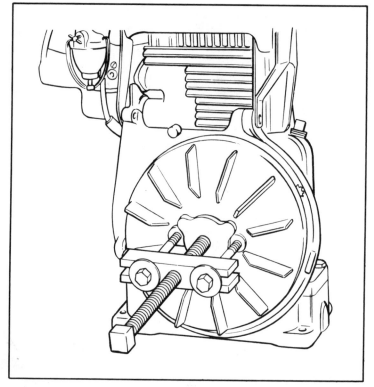

Fig. 2-15. A flywheel puller for most American engines can be fabricated or ordered from Kohler (shown) and other manufacturers. Briggs & Stratton and most other engines do not have pre-tapped puller mounting holes. Self-tapping capscrews must be used. Small European and Japanese engines are a special problem because their flywheels commonly use a threaded counterbore. Threads are metric and a special puller is required, available from the manufacturer or, as the case may have it, from bicycle dealers (these tools are also used to disassemble cotterless bicycle cranks.)

● Clinton and Tecumseh supply flywheel knockers (Tecumseh part Nos. 670103 for 7/16-inch crankshafts and 670189 for 1/2-inch shafts, both right-hand threaded). These tools dislodge the flywheel by impact and should be used with great care. A glancing blow can break or bend the crankshaft stub, and light, ineffectual blows can scramble flywheel magnets (Fig. 2-16).

● Although no factory recommends it and conscientious mechanics abhor the practice, flywheels have been removed with a brass bar used as a knocker. A large screwdriver used as a prybar behind the wheel helps reduce breakaway force, but is no guarantee that the crankshaft will not be ruined.

Inspect the flywheel hub for cracks (Fig. 2-17). Remove the key from the crankshaft, and replace if worn or distorted (Fig. 2-18). Check both flywheel and crankshaft keyways for excessive clearance that will allow the crankshaft to move relative to the flywheel. Movement on the order of a few hundredths of an inch as measured at the flywheel rim will upset edge distance and could result in hard starting. Unfortunately, there is no way to restore keyways, and damaged parts should be replaced.

It is not necessary to finally assemble the flywheel to test magneto output. Replace the key, slip the flywheel over the crankshaft and—with spark plug removed—hold the high tension lead with its bared output terminal 1/8 inch or so from the block. Spin the flywheel by hand; be careful not to cut your fingers on the governor air vane or other obstructions near the flywheel rim. A spark should be generated.

Once satisfied that the system operates, check key position as shown in Fig. 2-19. Replace the lockwasher and tighten the flywheel nut to manufacturer's specifications.

Contact Points

Small engine point assemblies are built in two basic

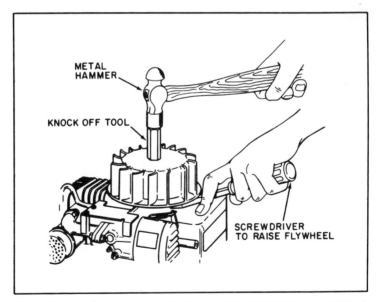

Fig. 2-16. If you use a knocker, be certain it matches properly with the crankshaft threads.

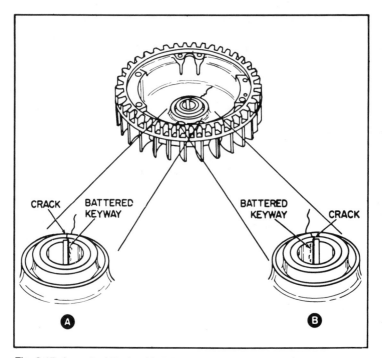

Fig. 2-17. A cracked flywheel hub is bad news, but there is some consolation in knowing how the damage occurred. A crack on the trailing side of the keyway means impact damage (A). The crankshaft suddenly stopped and the flywheel attempted to overtake it. A crack on the leading side of the keyway (B) means that the crankshaft was overspeeding the flywheel and can only occur if the flywheel nut were loose.

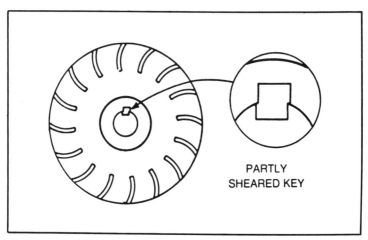

Fig. 2-18. Damaged flywheel keys must be replaced.

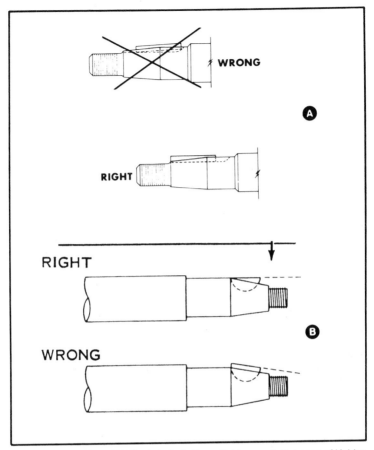

Fig. 2-19. When there is latitude in installing a flat key, as in the case of Kohler engines, position the key flush with the crankshaft shoulder (A). Half-moon keys are installed parallel to the crankshaft centerline (B).

configurations. What we might call the "standard" configuration consists of a pivoted point arm, a flat point spring, and a fixed arm (Fig. 2-20). Contacts are discs of tungsten (that can be perforated for cooling). The pivoted arm and its spring are electrically "hot."

The pivoted arm may bear directly against the cam and often features a nylon or phonelic rubbing block at the cam interface for better wear resistance and gap stability. Alternately, the point assembly can be mounted at some distance from the cam and articulated by means of a plunger.

The other configuration has been used on Briggs & Stratton under-flywheel magnetos for many years. In this design, the fixed

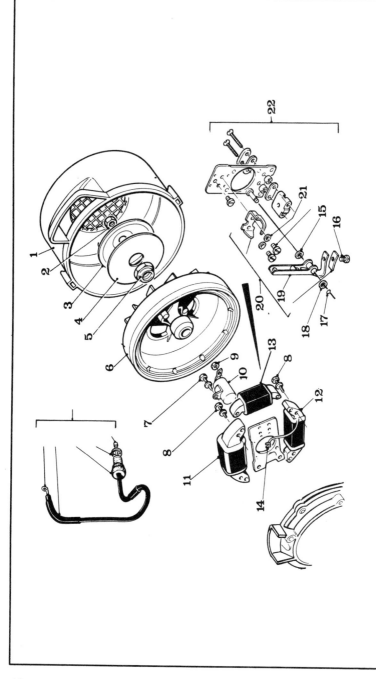

INDEX	PART NUMBER	QTY ASSY	REF	DESCRIPTION
1	11111 18 000	1		Flywheel Cover
2	11797 00 000	1		Nut
3	11794 00 000	1		Rotor Deflector
4	11795 00 000	1		Rubber Gasket
5	11796 00 000	1		Lock Nut Rotor
6	12131 00 000	1		Rotor
7	12703 00 000	1		Bolt
8	11081 00 000	6		Bolt
9	00056 00 000	1		Nut D4/75 - H35
10	11191 00 000	1		Condenser
11	11117 00 000	1		Ignition Coil
12	13294 00 000	1		Lighting Coil
13	13309 00 000	1		Stoplight Coil
14	00925 00 000	1		Lighting Coil Terminal
15	11218 00 000	1		Washer
16	00959 00 000	1		Bolt
17	11220 00 000	1		Circlip
18	11219 00 000	1		Washer
19	12584 00 000	1		Points Lever - Complete
20	11947 00 000	1		Contact (Kit 3)
21	00339 00 000	2		Washer
22	11120 00 000	1		Points

Fig. 2-20. Velosolex under flywheel magneto showing standard configuration point set (20). The hooked bracket mounts the fixed, or ground-ed, contact.

point is integral with the condenser and is hot. The grounded arm swings on a kind of rudimentary hinge and is secured by a small coil spring.

Troubleshooting. Points and condenser are sacrificial items and should be replaced at the first sign of ignition trouble that a spark-plug change does not cure. In rough order of frequency, point failure is occasioned by:

• High resistance across the contacts, caused by oxidation and/or metal migration. Normal contacts are slate gray in appearance, rough but without the peak and valley associated with metal transfer. Oxidized points are dark (sometimes black). The movable arm might show blue temper stain from overheating.

• Loss of point gap caused by wear on the plunger or rubbing block.

If contacts are good, points can be reset and tested.

• Loss of spring tension usually associated with point oxidation and subsequent overheating, but may be the result of fatigue. Symptom is high speed misfiring as points bounce and float. Because small engine manufacturers do not supply spring tension data, test by replacement.

• Oil fouling from seal failure is recognized by a splatter of carburized oil under the point contacts. Main bearing or plunger seal is involved.

• Broken or frayed primary wiring is almost always the fault of a mechanic who misrouted the wire.

Servicing. "STANDARD" configuration points are secured to the baseplate with one or two screws and can be located by short pins. When working quarters are tight—as in the case of Bosch-pattern magnetos that are serviced through small flywheel "windows"—it is advisable to use a magnetized screwdriver. The electrical connection between the moveable arm and the primary circuit usually takes the form of a small screw, but some point sets use the stationary point bracket as a tie point and incorporate an insulator (Figs. 2-21).

Upon removal of the point assembly, remove all traces of oil from the mounting area. Lubricate the cam with high-melting point grease and, when present, oil the cam wick. The point pivot can be oiled.

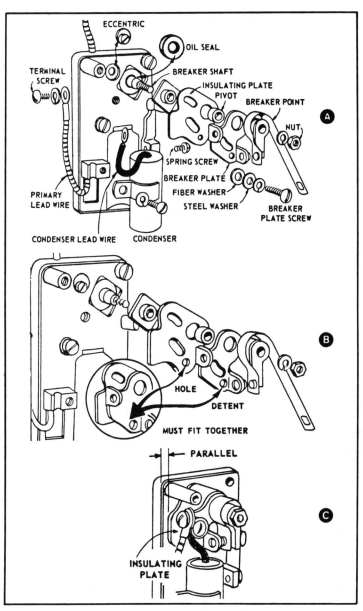

Fig. 2-21. Briggs & Stratton manufacturers these remotely mounted, standard configuration points as part of their Magna-Matic ignition system (A). Point sets can be complicated. Make careful note of the parts layout. The detent in stationary arm bracket must be indexed with a hole in the insulating plate (B) and, upon assembly, the breaker plate must be parallel with the left-hand edge of the breaker box (C).

49

Install the new point set, aligning pins are present with indexing holes. Tighten the electrical connection; be careful not to twist the moveable arm spring into contact with ground. Lightly secure hold-down screw(s).

While few mechanics take time to verify that contacts are parallel and concentric, this step can extend point life. While A of Fig. 2-22 illustrates the correct contact pattern, B of Fig. 2-22 shows the loss of contact surface that results from misalignment. Correct by bending the *fixed* contact with hold down screws *tight*. A bending bar, such as the one supplied by Tecumseh, should be used to avoid scratching the contacts. Use the following procedures to adjust the point gap:

Step 1. Made a preliminary adjustment so that points make and break as the flywheel is turned. Lightly snug the hold-down screw.

Step 2. Turn the flywheel until points open to maximum extension. When possible, visually verify that the high point of the cam is against the rubbing back or point-actuating plunger.

Step 3. As mentioned earlier, 0.020 inch is the more or less universal standard point gap for small engines, specified for most

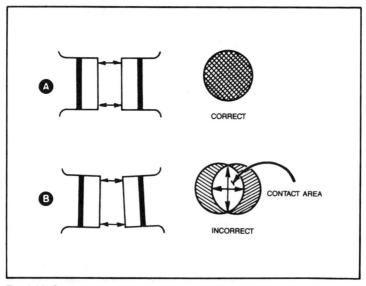

Fig. 2-22. Correct point alignment (A) results in full contact and maximum service life. Lack of parallelism or eccentricity reduces contact area (B). While visual examination is usually sufficient to determine alignment, in restricted quarters contact pattern can be registered by closing the points on a piece of plain paper with carbon paper.

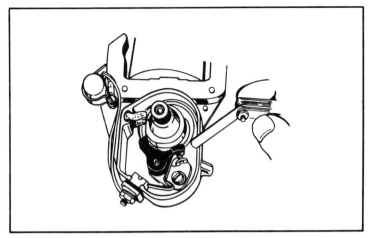

Fig. 2-23. Point gap adjustment on under-flywheel magneto.

and runable in any. This gap can be set with a 0.020-inch feeler gauge or, perhaps more accurately, with 0.019-inch and 0.021-inch blades used as go/no-go gauge.

Step 4. Using the screw or screwdriver-pry slots provided, move the fixed contact until the 0.019-inch go gauge slips easily between them and the 0.0210-inch no-go gauge forces the moveable contact open against its spring (Fig. 2-23). Gauge blades must be held dead parallel to contact surfaces during this operation.

Step 5. Tighten the hold-down screw(s) and recheck the gap. Distortion as the assembly is tighted usually affects gap and the adjustment must be repeated (this time anticipating for the change).

Step 6. Burnish points with a piece of cardboard (torn from the box that the points came in) to remove fingerprints, oil from the feeler gauge, and oxidation (Fig. 2-24).

Replace the condenser with the points, cleaning the condenser mounting area and routing the lead wire clear of the flywheel and other moving parts. Replacement procedure for integral Briggs & Stratton point and condenser sets is a follows:

Step 1. Remove the dust cover, held by two 1/4-inch self-tapping screws.

Step 2. Remove the condenser hold-down and point-support screws (Fig. 2-25).

Step 3. Disengage the coil and optional kill switch wires from the condenser, using the tool supplied in the Briggs & Stratton replacement point-set package or, lacking that, by "unscrewing" the spring with a pair of miniature water pump pliers.

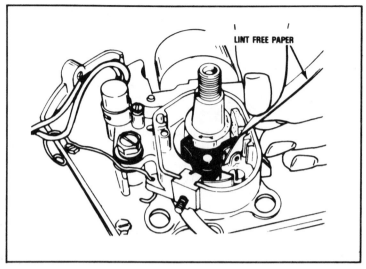

Fig. 2-24. After adjustment, point contacts should be burnished with paper to remove oil and possible oxidation. Snap points to remove any lint.

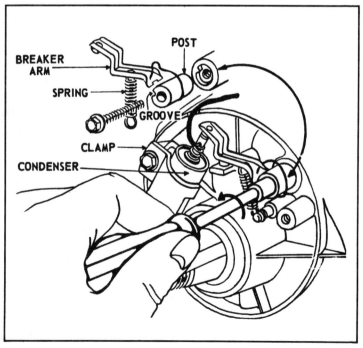

Fig. 2-25. Point disassembly for Briggs & Stratton light and medium frame engines. Note how the ground wire on the moveable point arm is routed.

Step 4. Inspect for oil in the point cavity; replace crankshaft seal as necessary.

Step 5. Inspect the point plunger for wobble. The plunger hole can be reamed and bushed, if necessary.

Step 6. Replace plunger if worn down to 0.870 of an inch or less (A of Fig. 2-26). Assemble with the grooved end toward the points (B of Fig. 2-26).

Step 7. Install the replacement moveable points arm by first engaging the post on its mount, mating the slot in the post with the indexing tab.

Step 8. Tighten the point hold-down screw.

Step 9. Position the moveable point arm for installation with its braided ground wire outboard of the post, as shown in Fig. 2-27.

Step 10. One end of the point spring is open, the other end forms a closed loop. Guide the open end through two holes provided in the moveable point arm. Slip the closed end of the spring over the groove in the spring mounting post.

Step 11. Grasp the moveable arm and, pulling against spring tension, engage it into the slot provided in its mounting pedestal.

Step 12. Using the depressor tool, install the coil and kill switch wires on the condenser terminal (Fig. 2-28).

Step 13. Rotate the crankshaft to retract the point plunger.

Step 14. Install the condenser so that point contacts lightly touch and snug the hold-down screw.

Step 15. Turn the crankshaft to extend the plunger and open the points.

Step 16. Measure the gap with a 0.020-inch feeler gauge, moving the condenser as required to establish the correct gap.

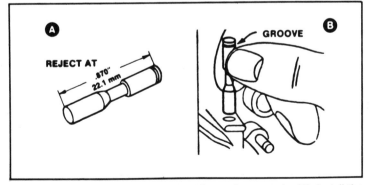

Fig. 2-26. Excessive plunger wear can affect point geometry (A). Install the plunger with the groove adjacent to the moveable point arm (B).

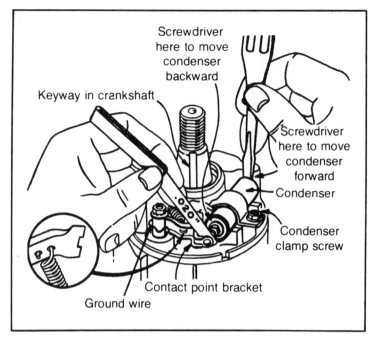

Fig. 2-27. Gap points with braided ground wire outboard of point arm pedestal and open eyelet of spring hooked to moveable point arm. Turn the crankshaft to bring the keyway adjacent to the point plunger; snug the clamp screw and set the point gap 0.020 of an inch. Use a screwdriver to move the condenser into or away from the moveable arm contact. Tighten clamp screw and check gap.

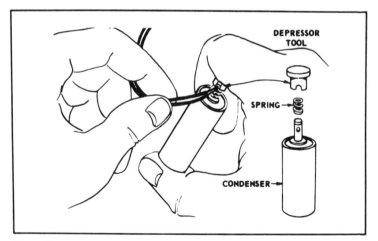

Fig. 2-28. A depressor tool, packed in Briggs & Stratton replacement point sets, should be part of a small engine mechanic's tool kit.

Step 17. Tighten the condenser hold-down screw and recheck the gap. Some "creep" is inevitable and the operation will have to be repeated.

Step 18. Burnish the contacts with cardboard.

Step 19. Install the flywheel key and flywheel.

Step 20. Test magneto output by spinning flywheel.

Filing Points

Ideally, burnt contact points should be replaced—together with the offending condenser—but this is not always practical in the field. Do what you can. Use a fine, single-cut file and work patiently until both contacts are bright. Some mechanics mount the point assembly in a vise and dress the contacts slightly convex to assure that they meet over a wide area. But, in an emergency, all that's needed is to close the contacts over a file and get after it.

Ignition Coils

While ignition coils can be tested with the proper equipment, substituting a known good coil is the best test. Battery and coil systems are tolerant and any coil of the same rated voltage can be used as a test. Indeed, it is common practice to permanently substitute department-store specials for expensive original equipment manufacture (OEM) coils on some Japanese motorcycles.

Unfortunately, magneto coils offer few opportunities for interchange (although it is possible to swap coils from other engines that use the same, or a similar, magneto). Older designs have the coil secured to the armature by means of a spring wire clip, and it is sometimes possible to replace the coil without disturbing armature hold-down fasteners. More modern designs usually integrate coil and armature into a single assembly.

Whenever the armature is disturbed on a magneto system, it is necessary to re-establish armature-to-flywheel (or rotor) distance. This distance varies with manufacturer and model, but is usually in the neighborhood of 0.006 of an inch to compensate for manufacturing tolerances, worn bearings, and thermal expansion. The traditional way to establish the air gap is to insert shim stock of required thickness between the flywheel (or rotor) and the loosely attached coil assembly (Fig. 2-29). Turn the crankshaft until the magnets are under the armature, tighten the armature hold-downs and remove the shim stock. A business card can be substituted for shim stock at some sacrifice in precision.

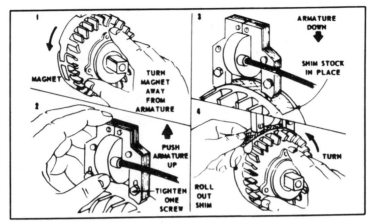

Fig. 2-29. Briggs & Stratton coils mount outside the flywheel, simplifying air gap adjustment.

Phelon, Wico, and a number of other magnetos are completely contained in the flywheel cavity and do not have built-in provision for adjustment. Normally, the air gap is ignored, on the theory of letting sleeping dogs lie, but an inquisitive mechanic can determine the gap by wrapping the armature with vinyl electrical tape (Fig. 2-30). Two layers of tape, applied without creases and without

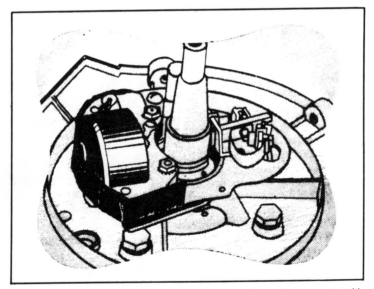

Fig. 2-30. Vinyl electrical tape serves as an air gap indicator for magnetos with under-flywheel coils. Clinton shown.

stretching for a total thickness of 0.014-0.018 of an inch, should cause interference when the flywheel is torqued down and rotated. (Because of tolerance stack and greater heating effects, the under-flywheel-coil air gap is about twice that of coils mounted outboard of the flywheel.) The air gap can be reduced by judicious parts changing and can be increased by filing the ends of the armature. Nevertheless, armature ends must be filed evenly, because each leg of the laminations must be the same distance from the magnets.

Edge distance, E-gap, breakaway gap, and pole shoe break describe the spacial relationship between the ignition coil armature and the magneto magnet at the moment of point break. This relationship is fixed on American engines, or, if not, is clearly marked (Fig. 2-31). On some foreign engines pole shoe break must be re-established whenever ignition coil hold-down screws have been disturbed. Figure 2-32 illustrates pole shoe break for a high-performance Bosch magneto. Other manufacturers measure from different referents.

Optimum pole shoe break can be approximated with the aid of point gap variations. Increasing the gap narrows pole shoe break. Thus, if an engine has a point gap specification of 0.020 inch and produces its best spark at, say, 0.026 of an inch, the pole-shoe break should be increased as necessary to give best spark at specified point gap.

SOLID-STATE SYSTEMS

Capacitive discharge ignition (CDI) systems have all but made

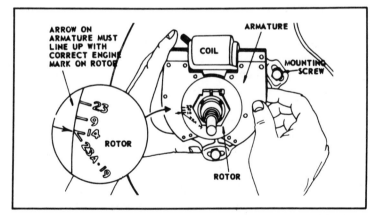

Fig. 2-31. Briggs & Stratton Magna-Matic E-gap varies with engine model; armature is marked accordingly.

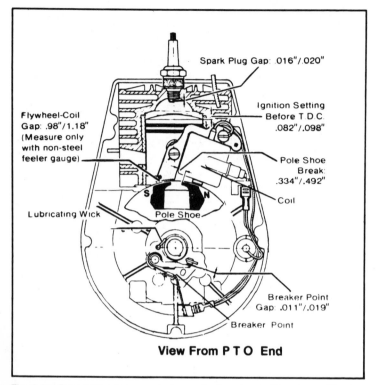

Spark Plug Gap: .016"/.020"

Ignition Setting
Before T.D.C.
.082"/.098"

Flywheel-Coil
Gap: .98"/1.18"
(Measure only
with non-steel
feeler gauge)

Pole Shoe
Break:
.334"/.492"

Coil

S   N

Lubricating Wick
Pole Shoe

Breaker Point
Gap: .011"/.019"

Breaker Point

View From P T O End

Fig. 2-32. Robert Bosch magneto has pole shoe break (E-gap) specification of 0.334-0.492 of an inch, measured between north pole lamination and coil housing.

conventional systems obsolete. While circuit details vary somewhat between makes and models, basic operation, shown schematically in Fig. 2-33, is similar. The flywheel magnet (1-A) passes the input coil, generating an ac voltage of about 200V at normal engine speeds. Rectifier (3) converts alternating current to direct current, suitable for storage in a large capacitor (4). About 180° of crankshaft rotation later, the magnet passes a trigger coil, generating a small voltage across resistor (6). This voltage causes the silicon-controlled rectifier (7) to become conductive, releasing the charge on the capacitor (4) into primary windings of the pulse transformer (8). A large voltage is generated in the transformer's secondary windings that goes to ground across the spark plug (9) electrodes. Some CDIs employ a second trigger coil to retard the spark at low rpm; others feature electronic advance (keyed to rpm as a function of trigger-coil voltage).

These systems are highly integrated and may combine the pulse transformer and input coil in the same housing. Low-voltage generating coils needed to provide power for accessories and battery charging may also be present.

CDI systems have three advantages over the systems they replace. Output voltage, particularly on outboard motor and snowmobile applications, is extremely high. Voltage buildup is rapid and carboned spark plugs usually will fire before voltage has time to leak to ground. And, perhaps most important, contact points are eliminated (together with the rationale for frequent tune-ups).

On the debit side, CDIs are not repairable in any significant sense. Other than to inspect wiring and remove rust accumulations on the magnet faces (that can cause high-speed misfire), there is little a mechanic can do.

Some systems are entirely encapsulated and must be changed out as a unit. Others are divided into low- and high-tension sections,

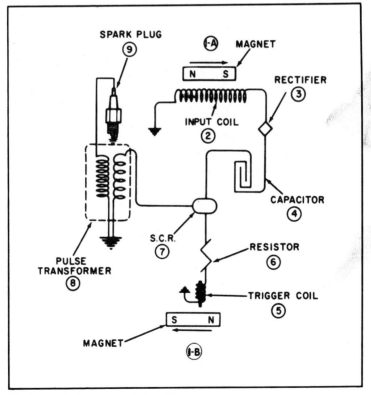

Fig. 2-33. The Tecumseh CDI system.

with the low-tension generating and switching circuits near the magnets and the high-tension coil mounted on the engine block or frame. Dealers test equipment will at least identify which half of the system should be replaced.

Other designs employ a discrete trigger coil with leads to the encapsulated remainder of the circuit. Trigger coil output can be checked with a very sensitive voltmeter or, lacking that, with an ordinary ohmmeter set on its lowest range. Connect the ohmmeter across trigger coil output leads and spin the flywheel. The meter will see coil output voltage as a sudden increase in resistance.

Some manufacturers have not completed the changeover to CDI, and otherwise identical engines come off the line with either conventional or solid-state ignitions. Magneto ignition coils on some Tecumseh engines even look like CDI modules made by the same company, but the magneto version requires a gray stepped flywheel key while the CDI uses a gold key (Fig. 2-34).

Magnetron Retrofit. Briggs & Stratton Magnetron CDI entered series production in August 1982, replacing the magneto shown in Figs. 2-25 through 2-28. The $15 Magnetron kit, part No. 394970, can be used to update ignition systems in older, single-cylinder, aluminum block engines with the "two-legged" armature (as shown in Fig. 2-29). In addition, the kit is the replacement part for failed Magnetron units on cast-iron block, single- and twin-cylinder engines. First-time installation requires about an hour.

Step 1. Disconnect the spark plug lead and remove the blower housing (flywheel shroud).

Step 2. The factory suggests that the flywheel be removed to access primary and kill-switch wiring, and to replace the flywheel key with an identical key supplied in the kit. Flywheel removal is not mandatory, because the wires can be snipped at their exit point at the dust cover (Fig. 2-35). However, the wires will be several inches short of the Magnetron and must be spliced.

Step 3. Normally the points and condenser are left undisturbed. If you want to discard the points, the point plunger hole must be sealed with part No. 231143 for single-cylinder engines or part No. 231262 for twins. Failure to block off the hole results in a serious oil leak.

Step 4. Remove the two capscrews holding the ignition coil armature and governor vane bracket to the engine block. *Note*: Four rivet heads identify the visible side of the armature (Fig. 2-35). Some models require disengagement of the air vane from its wire and spring linkage.

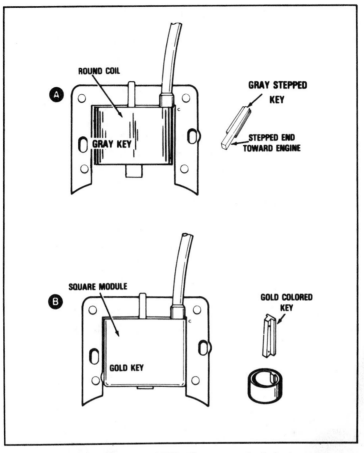

Fig. 2-34. Tecumseh magneto and CDI coils are vaguely similar in appearance, but not interchangeable. The magneto coil is cylinderical and used in conjunction with a gray flywheel key (A). The CDI unit is rectangular and requires a gold key (B).

Step 5. Hold the ignition coil armature upright, standing it on its "legs" with its inner side (the side without rivet heads) facing you. As the part is now oriented, the Magnetron module installs between the coil and the *right* leg.

Step 6. Using finger pressure only, slip the module into position, making certain its plastic hook locks on the armature shoulder. See Fig. 2-36.

Step 7. Prepare to make the electrical hookup. One or two hot (insulated) wires ran to the ignition points. One wire connected the points with the coil primary and the other, optional, wire went to the kill switch, located near the carburetor. The ignition coil was also grounded through a bare wire that attached to one of the coil armature hold-down screws. The Magnetron module has two wires: a long ground wire with a terminal on its end and a short, uninsulated hot wire.

Step 8. Remove the insulation about 3/4 of an inch back from the wire(s) that originally went to the ignition points. Scrape bare wire ends to remove varnish.

Step 9. Route the wire(s) in under the ignition coil to lay between the coil and cylinder upon assembly. If the flywheel has been removed, wire length should be sufficient to reach the Magnetron. If not, splice in additional wire, twisting the ends for mechanical strength and soldering with 60-40 rosin core solder. Do not use crimp-on connectors for these or other connections. Complete the job with heat-shrink tubing or vinyl electrician's tape.

Step 10. Connect these wires to the short, uninsulated Magnetron hot wire. Twist the wires, making two full turns, and secure with a small amount of rosin-core solder, using only enough heat to flow the solder into the joint (Fig. 2-37).

Step 11. The installation kit contains a small coil spring and a J-shaped connector. Slip the coil spring over the long arm of the

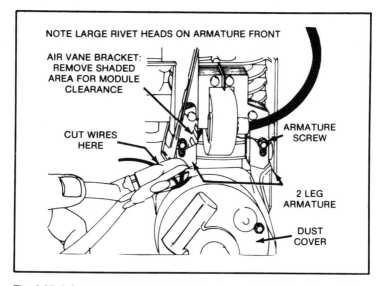

Fig. 2-35. It is not absolutely necessary to remove the flywheel to retrofit a Magnetron CDI on many Briggs & Stratton engines. However, flywheel removal simplifies the work. Note the four rivet heads visible on coil laminations that reference the outboard (flywheel) side of laminations.

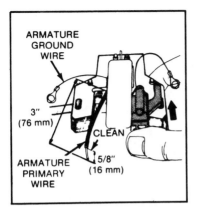

Fig. 2-36. The Magnetron module must be installed on right-hand coil lamination "leg" as viewed from the underside of the armature/coil assembly. In this position, rivet heads will not be seen.

J so that its end abuts the crossbar. This assembly is shown in Fig. 2-38. Insert the J connector—hook first—into the hole provided on the side of the module. Using a drill bit, punch or the contact-point end of a Briggs & Stratton condenser, press the connector into the bore against spring tension. The connector might have to be turned to align the hook with the slot in the bore. Slip the wires soldered in the previous step under the connector hook and withdraw the tool to release spring tension. The connector will retract (holding the wires securely). Snip off the ends of the wires 3.16 of an inch beyond the connector.

Step 12. Twist the armature ground and module ground wires together close to the armature (Fig. 2-39). Solder the connection.

Fig. 2-37. Coil hot (insulated) wire(s) are twisted together and soldered to the short, uninsulated Magnetron hot wire and installed in the unit with a special clip.

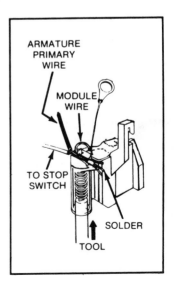

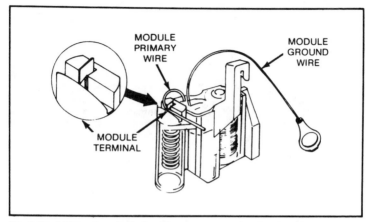

MODULE
PRIMARY
WIRE

MODULE
GROUND
WIRE

MODULE
TERMINAL

Fig. 2-38. A Magnetron J-clip (inset) secures a previously soldered coil hot and Magnetron hot wires to the module body.

Because both wires have terminals, the shorter of the two can be snipped off.

Step 13. To prevent vibration damage, secure the wires under the coil with Permatex or another sealant.

Step 14. Install the armature, ground wire, and governor vane. The ground wire is not to be attached on the air vane side of armature. Armature-to-flywheel clearance is 0.006 to 0.010 of an inch for engines up to 13 cubic inches displacement and 0.010 to 0.014 of an inch for large models. (The first two digits in the model

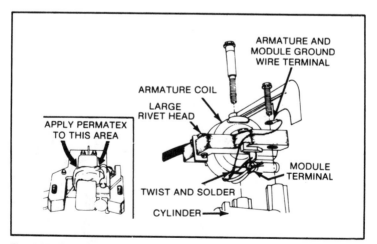

ARMATURE AND
MODULE GROUND
WIRE TERMINAL

ARMATURE COIL

LARGE
RIVET HEAD

APPLY PERMATEX
TO THIS AREA

MODULE
TERMINAL

TWIST AND SOLDER

CYLINDER

Fig. 2-39. The unit is installed on the block with large rivet heads up, wires are secured with Permatex, and the coil air gap is adjusted.

number indicates cubic inch displacement—60000 translates as 6.0 cubic inches and 140000 means 14 cubic inches).

Step 15. Test the spark output by spinning the flywheel at least 350 rpm.

Step 16. Complete assembly and the start engine. If the system produces spark but the engines does not start, suspect a sheared key.

Chapter 3

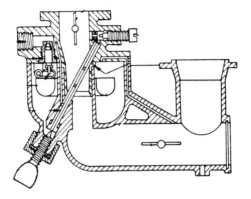

Carburetors and Fuel Systems

The carburetor is a much-maligned instrument whose primary purpose is to atomize liquid gasoline into an explosive mist. Unfortunately for mechanics, this is not as easy as it sounds and the hardware can be a bit intimidating.

OPERATION

Carburetors have five basic elements. The *throttle*, which is part of the carburetor closest to the engine, is a valve that determines how much air-fuel mixture is delivered. On American designs, the throttle is a pivoted plate, mounted across the carburetor bore (Fig. 3-1). European and some Japanese manufacturers favor a sliding throttle, or shuttle, that pulls upward out of the carburetor bore. All throttles incorporate an adjustable stop on the closed side that determines engine rpm at idle.

Gasoline moves through the carburetor along two channels, called circuits, in the carburetor body casting. The *low-speed circuit* delivers fuel through one or more tiny holes just upstream of the throttle. As its name indicates, the low-speed circuit functions when the throttle is nearly closed. An adjustment needle is provided to regulate the amount of gasoline or, in a few examples, the amount of air in the mixture. The *high-speed circuit* empties through a port or removable jet located upstream of the low-speed port. It comes into play when the throttle is opened, although there is an overlap

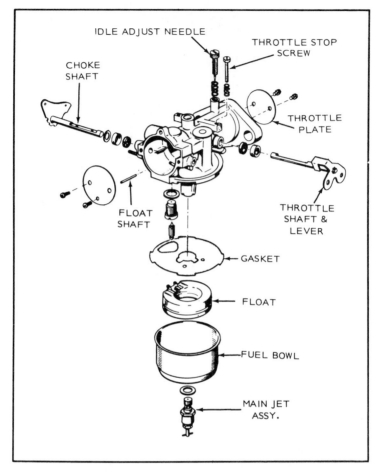

Fig. 3-1. Basic carburetor nomenclature. The example shown is an Onan float-type, side-draft carburetor using a pivoted throttle plate. An area of confusion for beginning mechanics is that nomenclature is not entirely consistent among manufacturers. For example, the throttle plate is called the throttle valve or butterfly; the throttle stop screw is also known as the idle rpm screw; the idle adjust screw is termed the idle, or low speed, adjustment; and the adjustment on the main jet assembly in this drawing, is labeled the power, or high speed, adjustment, main adjust needle, or needle valve.

period when both circuits flow. A high-speed mixture adjusting needle is almost always present.

Because large amounts of fuel are needed during cold starts, carburetors incorporate a starting system to enrich the mixture. In most instances, a choke plate is mounted just behind the air cleaner. When closed, the choke blocks air entry into the carburetor bore and, at the same time, encourages fuel flow by subjecting high-speed and low-speed circuits to a partial vacuum. Some foreign designs use a starting jet, which is merely a large orifice that is opened to provide additional fuel. A few American carburetors use a primer pump to squirt raw gasoline into the cylinder for starting.

Whether supplied by gravity or a fuel pump, more gasoline is available to the carburetor than the engine can burn. Consequently, carburetors employ some means of internal fuel regulation. Most often this is achieved by a float-controlled inlet valve—reminiscent of a toilet tank float—that opens to admit fuel as the engine requires it. In some instances, a diaphragm—triggered by pressure fluxuations in the engine crankcase—substitutes for the float. But not all carburetor diaphragms are used to regulate internal fuel levels. In many cases the diaphragm is the active element of a fuel pump and a float or a second diaphragm regulates the carburetor fuel level.

TYPES

Carburetors can be conveniently divided into three groups on the basis of hardware used to maintain internal fuel level.

Suction-lift

Single-cylinder utility engines are often fitted with suction-lift carburetors that are mounted on the fuel tank and pick up fuel through a standpipe. Figures 3-2 and 3-3 illustrate Briggs & Stratton, and Clinton types.

Regardless of manufacturer, these carburetors are similar in design and can develop similar problems, including:

● Loose flange screws. The carburetor and gasoline tank is secured to the engine with two screws that can vibrate loose and cause an air leak at the flange gasket. If the problem is chronic, replace the flange gasket and secure the screws with Loc-tite or a similar product.

● Leaks at the fuel tank/carburetor body interface. Remove the tank, check for cracks radiating out from the screw holes, and replace as necessary.

Warning: Do not attempt to solder or otherwise repair this or any other fuel tank.

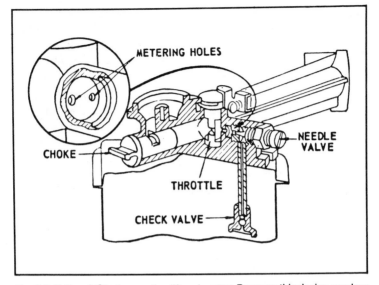

Fig. 3-2. Briggs & Stratton suction lift carburetor. Because this design employs a single mixture adjustment (labeled "needle valve"), the mixture tends to be slightly rich when adjusted for maximum power and smooth acceleration. In addition, this and other suction-lift carburetors may "hunt" when operated under no load at light and mid throttle.

● Sticking choke—corrosion and carbon buildup can cause Briggs & Stratton plug-type chokes to bind. Free with Gum-out or an equivalent product.

● Stuck check ball—the result of stale fuel. Replace the inlet pipe, twisting it counterclockwise to remove. Install a new pipe and integral check valve to the same depth as the original.

● Loose expansion plug (Clinton)—replace plastic plug.

Generally, the carburetor can be cleaned on the engine by removing the needle valve, spring, lock nut and o-rings. The jet will now be visible and can be opened with compressed air or with a soft bristle. Immersion in commercial carburetor solvent is not recommended.

Suction Lift (Fuel Pump)

Briggs & Stratton took suction lift technology a step further with the development of the "Pulsa-Jet" (Fig. 3-4). The carburetor feeds from a small chamber in the top of the fuel tank, which is replenished by the pump pipe. In this way, fuel available to the

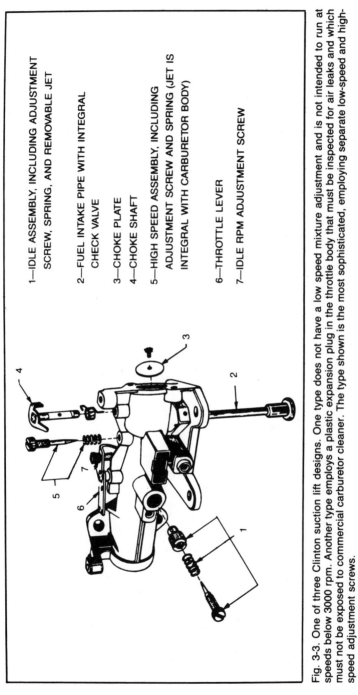

1—IDLE ASSEMBLY, INCLUDING ADJUSTMENT
SCREW, SPRING, AND REMOVABLE JET

2—FUEL INTAKE PIPE WITH INTEGRAL
CHECK VALVE

3—CHOKE PLATE

4—CHOKE SHAFT

5—HIGH SPEED ASSEMBLY, INCLUDING
ADJUSTMENT SCREW AND SPRING (JET IS
INTEGRAL WITH CARBURETOR BODY)

6—THROTTLE LEVER

7—IDLE RPM ADJUSTMENT SCREW

Fig. 3-3. One of three Clinton suction lift designs. One type does not have a low speed mixture adjustment and is not intended to run at speeds below 3000 rpm. Another type employs a plastic expansion plug in the throttle body that must be inspected for air leaks and which must not be exposed to commercial carburetor cleaner. The type shown is the most sophisticated, employing separate low-speed and high-speed adjustment screws.

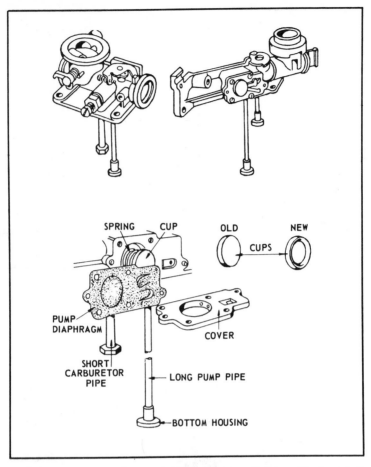

Fig. 3-4. The "Pulsa-Jet", a suction-lift carburetor with integral fuel pump, is manufactured in two styles. Note the elegant simplicity of design that makes it the most reliable of small-engine carburetors.

carburetor is always at a constant level. There are three things to remember about these carburetors:

● Unless the fuel tank is topped up, the engine will have to be cranked a few times on initial startup to fill the carburetor feed chamber.

● Failure to deliver fuel can almost always be cured by replacing the diaphragm.

● Replace any old style cups that might be encountered with the new type shown in Fig. 3-4.

Float Types

Float carburetors are by far the most popular type and are found on a wide variety of utility and high-performance engines. The ma-

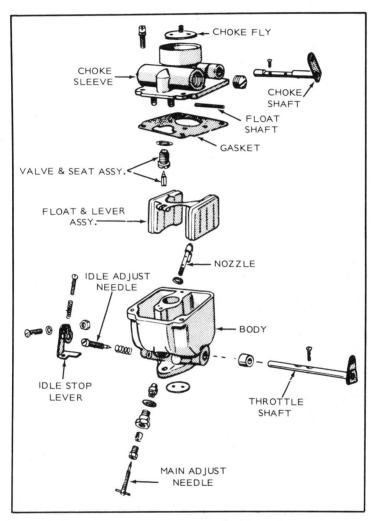

Fig. 3-5. Downdraft carburetor used on Onan CCK/CCKA twin-cylinder engines. As is traditional with downdraft designs, the main jet empties into a nozzle that lies athwart the air stream. Holes in nozzle aerate fuel before discharge for more consistent vaporization and better response to sudden throttle opening. Throttle shaft bushing indicates that the carburetor is intended for long service and repairability. Throttle shaft and bushing should be replaced when side play exceeds 0.008-0.010 of an inch.

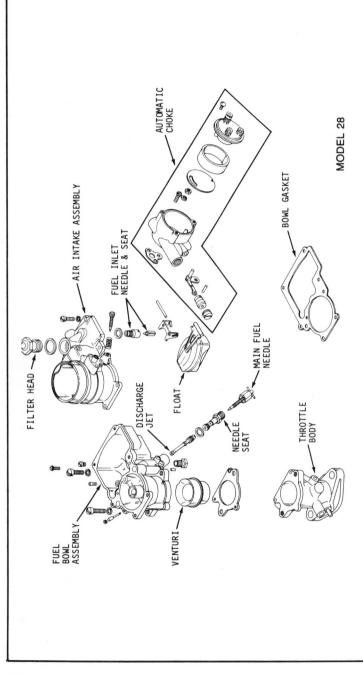

FILTER HEAD

AIR INTAKE ASSEMBLY

FUEL INLET
NEEDLE & SEAT

AUTOMATIC
CHOKE

DISCHARGE
JET

FLOAT

MAIN FUEL
NEEDLE

BOWL GASKET

FUEL
BOWL
ASSEMBLY

VENTURI

NEEDLE
SEAT

THROTTLE
BODY

MODEL 28

74

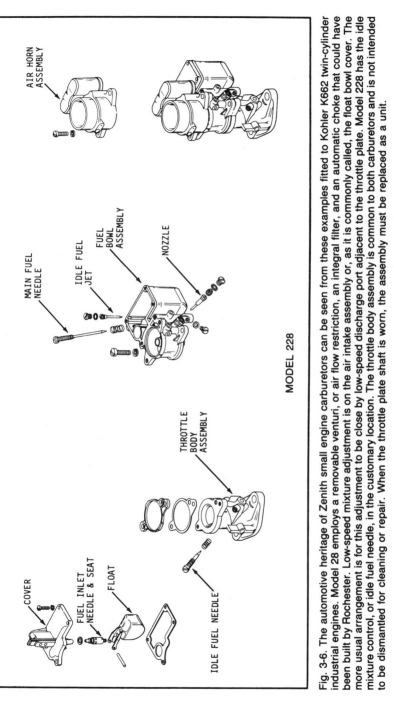

MODEL 228

Fig. 3-6. The automotive heritage of Zenith small engine carburetors can be seen from these examples fitted to Kohler K662 twin-cylinder industrial engines. Model 28 employs a removable venturi, or air flow restriction, an integral filter, and an automatic choke that could have been built by Rochester. Low-speed mixture adjustment is on the air intake assembly or, as it is commonly called, the float bowl cover. The more usual arrangement is for this adjustment to be close by low-speed discharge port adjacent to the throttle plate. Model 228 has the idle mixture control, or idle fuel needle, in the customary location. The throttle body assembly is common to both carburetors and is not intended to be dismantled for cleaning or repair. When the throttle plate shaft is worn, the assembly must be replaced as a unit.

75

jor limitation of this design is that the float chamber must be kept approximately level. Consequently, these carburetors are not used on chain saws, snowmobiles, and other equipment subject to changes in attitude.

Float-type carburetors fall into three classifications on the basis of mixture delivery. *Downdraft* carburetors receive air from the top, mix fuel in the central section, and deliver the mixture through the mounting flange at the bottom (Figs. 3-5 and 3-6). While almost universal on automobiles, application of this design to small engines is limited to multicylinder, horizontally opposed types. *Updraft* carburetors reverse the architecture just described. Air enters from the bottom of the instrument, mixes with fuel in the central section, and is expelled through the top. Figure 3-7 shows a typical design in cutaway view.

The Flo-Jet is one of the most reliable small engine carburetors, but it does have certain idiosyncrasies. Unless the engine starts almost immediately, gasoline will dribble from the air cleaner mounting boss. This condition is normal and is not to be taken as prima facie evidence of severe flooding. The main jet adjustment, called "needle valve," is quite sensitive but response is slow. Make a small adjustment, turning the thumb screw a fraction of a turn

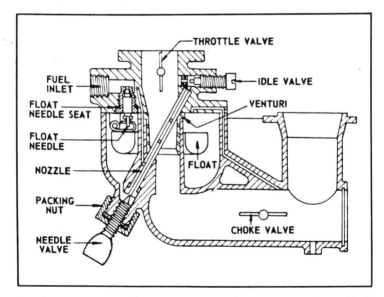

Fig. 3-7. Flo-Jet in cutaway view. In Briggs & Stratton nomenclature, this is a "two-piece" carburetor because it consists of two castings, the upper or throttle body and the lower or carburetor body.

and wait a few seconds for the instrument to respond. The fuel nozzle extends into the throttle body casting. Upon disassembly, the throttle body must be moved diagonally to disengage the nozzle. And finally, overtightening throttle body screws can distort the casting and cause fuel leaks around the gasket at the top of the float chamber. Correct by removing the throttle body and carefully straightening the distorted corners with a brass pin or miniature ball-peen hammer.

Side-draft carburetors are by far the most popular with applications that range from vertical crankshaft lawnmowers to snowmobiles. Parts arrangement is horizontal. Fuel mixing takes place in the central section above the float chamber. Figure 3-8 illustrates a typical example.

There are several things a mechanic should be aware of when dealing with the LMG and its kin: the float bowl gasket fit is crucial and new gaskets must usually be stretched like a rubber band before the gasket will cover the full diameter of the float bowl lip. Do not remove the fuel nozzle without a replacement part in hand. Walbro nozzles are cross-drilled after assembly to open a fuel passage to the idle circuit. It is almost impossible to realign this drilling-upon assembly and the engine, starved for fuel at low speed, will refuse to idle. Replacement nozzles, illustrated in the lower right of Fig. 3-8, have an angular groove that accommodates misalignment. The float drain may develop leaks and should be replaced as part of carburetor overhaul. If the bottom of the float bowl is rusted, however, the bowl must also be replaced.

Most side-draft carburetors use a pivoted throttle plate to regulate air-fuel delivery and engine speed. Some Japanese and European designs employ a throttle slide for the same purpose (Fig. 3-9). In the usual configuration, the throttle slide incorporates a tapered needle, which partially obstructs the main jet. As the slide retracts out of the carburetor bore to admit more air, the needle moves out of the jet, allowing more fuel to flow. At wide-open throttle, all that obstructs the carburetor bore is the needle.

Mechanics should remember that most piston slide carburetors regulate the low speed mixture strength by means of the air supply. Tightening the low speed mixture screw richens the mixture. The throttle piston is accessible after the piston cap and choke assembly are removed. Exercise care to see that dirt does not enter the piston bore and gently extract the piston. The metering pin is vulnerable and, once bent, will accelerate wear on the main nozzle. Lean running seems to be the main complaint with this type of

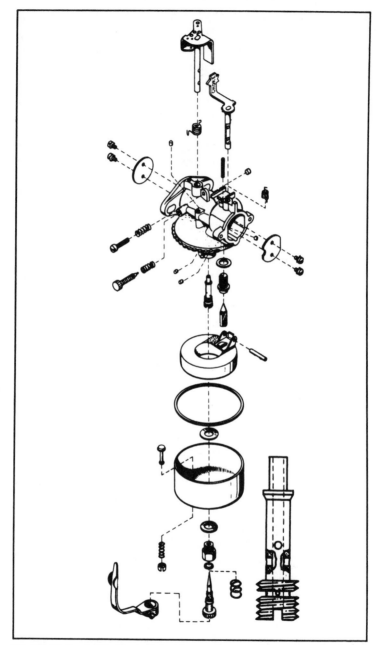

Fig. 3-8. Walbro LMG side draft, float-type carburetor. This very popular design is used by Clinton, together with almost identical LBM and LMV varieties, and by Tecumseh. (Courtesy Clinton Engines Corp.)

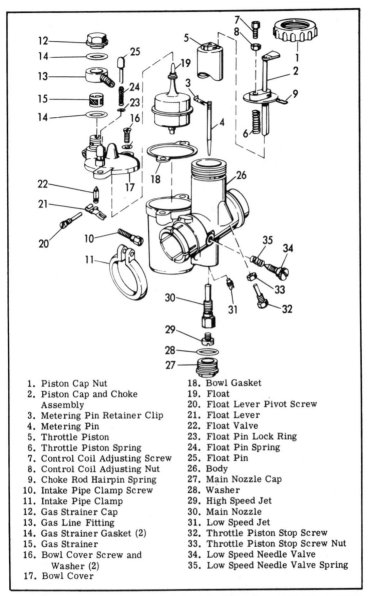

1. Piston Cap Nut
2. Piston Cap and Choke Assembly
3. Metering Pin Retainer Clip
4. Metering Pin
5. Throttle Piston
6. Throttle Piston Spring
7. Control Coil Adjusting Screw
8. Control Coil Adjusting Nut
9. Choke Rod Hairpin Spring
10. Intake Pipe Clamp Screw
11. Intake Pipe Clamp
12. Gas Strainer Cap
13. Gas Line Fitting
14. Gas Strainer Gasket (2)
15. Gas Strainer
16. Bowl Cover Screw and Washer (2)
17. Bowl Cover
18. Bowl Gasket
19. Float
20. Float Lever Pivot Screw
21. Float Lever
22. Float Valve
23. Float Pin Lock Ring
24. Float Pin Spring
25. Float Pin
26. Body
27. Main Nozzle Cap
28. Washer
29. High Speed Jet
30. Main Nozzle
31. Low Speed Jet
32. Throttle Piston Stop Screw
33. Throttle Piston Stop Screw Nut
34. Low Speed Needle Valve
35. Low Speed Needle Valve Spring

Fig. 3-9. This piston slide Del'Orto is typical of European motorbike practice and similar to a number of Japanese designs. The example shown was fitted to Harley-Davidson M-100 and M-125 bikes imported from Italy a few years back. The carburetor employs a cannister float, adjustable metering pin (or needle), replaceable high-speed and low-speed jets and a float pin. The last item, somewhat quaintly called a "tickler" by the English, is used to depress the float and flood the carburetor when other starting techniques fail.

carburetor, and on two-cycle applications, is often the fault of leaking crankshaft seals. Another possibility is extreme wear on the throttle piston and bore, which allows air to enter around the cap nut.

Float-type carburetors typically fail by flooding. This condition causes hard starting (if the engine is warm), and black smoke from the exhaust. In extreme and hazardous form, fuel escaping from the flooding results from:

- Wear on the inlet needle and seat.
- Dirt between the needle and seat (which sometimes can be dislodged by a sharp rap on the float cover).
- Improperly adjusted float level.
- Damaged float or float hinge.

Diaphragm Types

Diaphragm carburetors are conventional side-draft designs except that a diaphragm, rather than a float, is used to regulate the amount of fuel admitted to the instrument. The primary advantage of this arrangement is that the carburetor will continue to meter fuel when tilted at large angles off the horizontal. This is a useful attribute for chain saws, snowmobiles, motorbikes and, according to some designers, lawnmowers. Fuel cannot spill out of the instrument and, in absence of a float and float chamber, the design package is compact.

The crankcase is basically a sealed container. On four-cycle engines, it is partially filled with oil and somewhat tardily vented through the crankcase breather. Crankcase pressure fluxuates as the piston moves. On the downstroke pressure increases and on the upstroke it diminishes.

The frequency of pressure variations depends upon engine rpm and, as a consequence, can be used to control fuel delivery through the carburetor. Most diaphragm carburetors respond to negative, or low-pressure, crankcase pulses to momentarily open the fuel inlet valve. The type shown in Fig. 3-10 is typical of most applications. The more complex Tillotson model (shown in Fig. 3-11) employs both positive, or high pressure, and negative pulses. Positive pulses operate the fuel pump diaphragm on the bottom of the instrument that supplies fuel to the regulating diaphragm. The diaphragm then admits fuel to the carburetor circuits in response to negative pressure fluxuations.

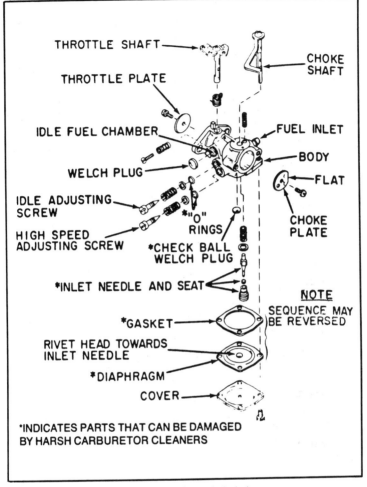

THROTTLE SHAFT

CHOKE SHAFT

THROTTLE PLATE

IDLE FUEL CHAMBER

FUEL INLET

BODY

WELCH PLUG

FLAT

IDLE ADJUSTING SCREW

*"O" RINGS

CHOKE PLATE

HIGH SPEED ADJUSTING SCREW

*CHECK BALL WELCH PLUG

*INLET NEEDLE AND SEAT

NOTE

SEQUENCE MAY BE REVERSED

*GASKET

RIVET HEAD TOWARDS INLET NEEDLE

*DIAPHRAGM

COVER

*INDICATES PARTS THAT CAN BE DAMAGED BY HARSH CARBURETOR CLEANERS

Fig. 3-10. The Tecumseh carburetor in exploded view.

The theory is a bit complex but a close look at Fig. 3-10 will help clarify matters. The neophrene diaphragm divides the diaphragm chamber into two parts. The upper part contains the inlet valve, which consists of a spring-loaded needle and seat. Fuel, impelled by gravity or by a remote pump, enters the upper half of the diaphragm chamber when the needle is raised. This part of the chamber also communicates to the crankcase by means of a port which passes through the carburetor body and mounting flange. The lower part of the chamber is dry and vented to the atmosphere by a small hole visible in the part labeled "cover."

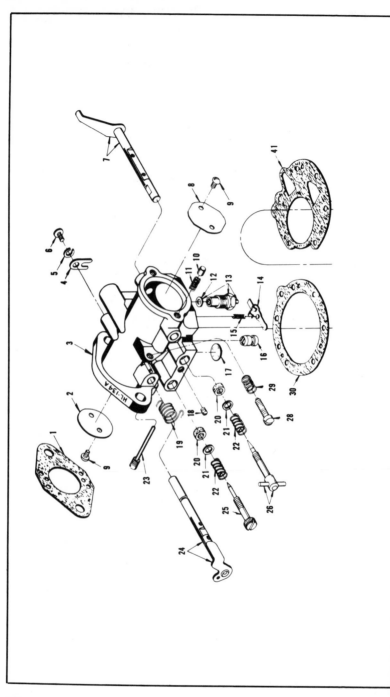

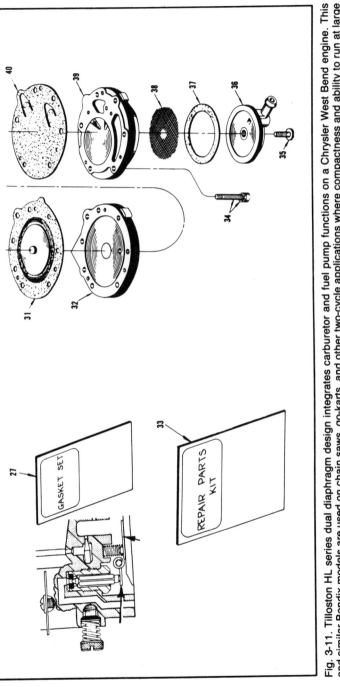

Fig. 3-11. Tilloston HL series dual diaphragm design integrates carburetor and fuel pump functions on a Chrysler West Bend engine. This and similar Bendix models are used on chain saws, go-karts, and other two-cycle applications where compactness and ability to run at large angles off the horizontal are important.

LLUS. NO.	QTY.	DESCRIPTION
1	1	Carburetor gasket
2	1	Throttle shutter
3		Order complete carburetor
4	1	Throttle shaft slip
5	1	Clip lockwasher
6	1	Clip retaining screw
7	1	Choke shaft and lever
8	1	Choke shutter
9	4	Shutter screw
10	1	Choke friction pin
11	1	Choke friction spring
12	1	Inlet seat gasket
13	1	Inlet needle, seat and gasket
14	1	Inlet control lever
15	1	Inlet tension spring
16	1	Nozzle check valve
17	1	Welch plug
18	1	Drain screw
19	1	Throttle shaft return spring
20	2	Adjustment screw packing
21	2	Adjustment screw washer
22	2	Adjustment screw spring
23	1	Control lever pinion screw
24	1	Throttle shaft and lever
25	1	Idle adjustment screw
26	1	Main adjustment screw
27	1	Gasket set
28	1	Idle speed regulating screw
29	1	Idle speed regulating screw spring
30	1	Diaphragm gasket
31	1	Diaphragm
32	1	Diaphragm cover
33	1	Repair parts kit
34	6	Body screw and lockwasher
35	1	Strainer cover screw
36	1	Strainer cover
37	1	Strainer cover gasket
38	1	Strainer screen
39	1	Fuel pump body
40	1	Fuel pump diaphragm
41	1	Fuel pump gasket
	1	Throttle shaft arm (not shown)

Parts listing for Fig. 3-11.

The ten commandments of this carburetor are: (1) replace the diaphragm and gasket after wet storage or whenever hard starting is a problem; (2) note that relationship between diaphragm and gasket shown in drawing is reversed on some models; (3) assemble diaphragm with rivet head—the round part, not the splayed shank—up; (4) inspect inlet needle and seat for contamination; (5) replace needle and seat in event of persistent flooding, visible wear or if needle tip is bent; (6) replace the seat with the aid of a six-point 9/32-inch socket, with its OD reduced by grinding; (7) some models incorporate a check ball in the high speed circuit.

If the ball and seat are replaceable, a welch plug will be present in the carburetor body above the diaphragm and adjacent to the high-speed adjustment screw; (8) inlet fitting is pressed into place and has an integral screen. If screen cannot be cleaned, extract fitting by holding it in a vise and cautiously twisting. Replace inlet fitting with a new part, positioning the fitting end for fuel line hookup; (9) atmospheric vent hole must be open in diaphragm cover; (10) wisdom dictates that harsh chemical cleaners should not be used because check ball, elastomer check valve in plastic inlet fitting (used on models with primer pumps) could be damaged.

When the engine is stopped, the diaphragm is relaxed and does not contact the extended needle tip. Consequently, the needle remains seated and no fuel flows past it. During cranking negative pressure pulses, generated on the piston upstrokes, stretch the diaphragm upward, unseating the needle. Fuel flows into the upper half of the diaphragm chamber and into the carburetor circuitry.

If for some reason the engine does not start, the diaphragm can be gently pushed upward by inserting a match stick through the cover vent. This unseats the needle and fills the upper part of the diaphragm chamber with gasoline.

Diaphragm carburetors arrived relatively late on the small-engine scene and share some of the characteristics of modern engineering. When they work, they work exceedingly well and allow a sharper tune and provide more power than most earlier types. But these carburetors are temperamental and demand a certain tolerance from mechanics. In rough order of frequency, problems are:

● Failed diaphragm, not always apparent under visual inspection. Replace diaphragm whenever hard starting or persistent fuel starvation occurs.
● Contamination by rust particles and other solids in the fuel,

which generally collects on the inlet screen (when fitted), needle and seat or in the low-speed circuit.

● Poor response to mixture adjustment controls, caused by forcibly seating adjustment screws into their jets. The Tillotson, whose operation is discussed above, can be assembled with high speed and idle adjusting screws interchanged. Screw tapers differ and the tip of the high-speed screw can break off in the low-speed jet. The inlet control lever, shown in Fig. 3-11 can be bent by rough handling and upset fuel metering.

Diaphragm carburetors are also adapted to Briggs & Stratton engines for junior-class kart racing, where power output is valued more than mechanical simplicity. From a mechanic's point of view, the carburetor diaphragm (31) is the most vulnerable part, followed by the inlet control lever (14). The lever should be replaced upon any sign of wear at the point where it touches the inlet needle. In addition, the lever must be parallel with the diaphragm housing as shown in the inset (Fig. 3-11). Correct any misalignment by carefully bending the inlet needle side of the lever while the lever is disassembled. Do not apply bending force to the needle that could deform the elastomer inlet seat and result in flooding. Nozzle check valve (16) malfunctions usually result in hesitant acceleration and poor mixture control. Draw down the six capscrews that secure fuel pump body (39) with extreme care, tightening in a criss-cross fashion. Failure to do this can wrinkle one or both diaphragms.

ADJUSTMENTS

High-speed mixture adjustment possibilities are as follows:

● Nonadjustable, nonreplaceable main jet, no adjustment possible.
● Nonadjustable, replaceable main jet, adjustment possible if alternate jets are available.
● Threaded adjustment needle, adjustment range from zero flow to flooded.
● Throttle slide adjustments, a special case discussed later.

Nonadjustable and nonreplaceable main jets are encountered, but most manufacturers make provision for the jet to be replaced with one of a larger or smaller orifice diameter. Replaceable main jets are commonly found on outboard motors to make adjustment possible while, at the same time, discouraging amateur attempts at tuning.

Threaded adjustment needles, used in conjunction with fixed or replaceable jets, are the norm and are illusrated throughout this chapter. The main jet needle, or screw as it is sometimes called, is usually located near the throttle plate on the carburetor body or else it is centered below the float bow.

Low-speed mixture adjustment devolves into these possibilities:

- Nonadjustable nonreplaceable jet, no adjustment.
- Threaded adjustment needle, with adjustment range from flooding to no flow.

Gasoline engines, particularity the small ones, do not like to idle. Very few carburetors have been made without provision for low-speed mixture adjustment. Most notable among these have been the sheet-metal Clinton suction-lift carburetor (for an engine that is not intended to operate below 3000 rpm), several versions of Tecumseh Craftsman carburetors, and the Lawnboy modular unit.

Other carburetors have such an adjustment that is almost always in the form of a tapered screw threaded into the carburetor body near the throttle plate. With exception of most piston slide carburetors, this screw controls fuel flow through the jet. Tightening the screw reduces the amount of fuel passing through the jet orifice and cleans the mixture.

Idle rpm adjustment is typically in the form of a spring-loaded screw that prevents the throttle from shutting off all fuel and air to the engine. Tightening the screw increases idle rpm.

Carburetor adjustment is a tune-up procedure and—unless the carburetor has been tampered with—should not be crucial for starting. The basic drill for diaphragm and float types with two mixture control screws and a conventional (pivoted) throttle is as follows:

1. Determine that the engine is in good tune, with ignition system functioning normally, diaphragm and float types, two mixture control screws, conventional, pivoted, throttle plate, and air filter services, and (when appropriate) float is set properly.

As shown in the Fig. 3-12, Tab A is the most crucial and

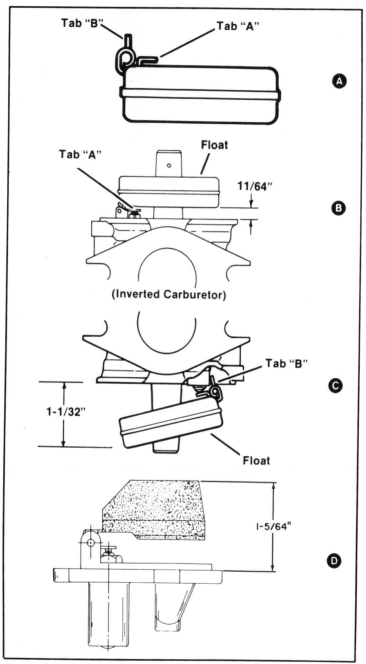

Fig. 3-12. Most floats incorporate two adjustment tabs.

88

controls float height. This adjustment is made by assembling float over needle and seat and inverting carburetor body (B). Exact adjustment varies carburetor make, model and application. However, as a general rule, the float should be parallel with carburetor body casting. Do not force the float into the needle and seat when bending tab. The second adjustment, controlled by tab B, is less critical (C). Lower edge of the float should not drop below main nozzle casting and should remain well clear of float bowl. Tillotson Model E float follows the general float height rule and is adjusted correctly when float parallels inverted carburetor body (D). Other carburetors employ an adjustment screw instead of tab A and some include a window or drain plug on float body so actual fuel level can be determined.

2. Crank the engine. If it does not start after several tries, lightly seat high-speed and low-speed mixture screws. Do not force these needles home: to do so will invite the kind of damage shown in Fig. 3-13 and destroy any hope of fine tuning. Back off both mixture controls about one full turn. The specification varies with carburetor type, but one turn from lightly seated should pass enough fuel for the engine to start.

3. Allow the engine to reach stable operating temperature. The choke must be full open.

4. With the throttle plate slightly more than half open, back off the high-speed mixture control screw in small increments,

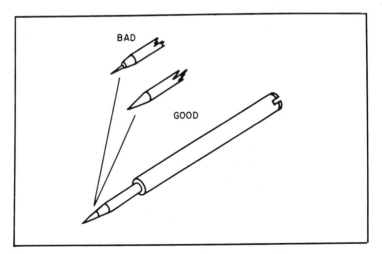

Fig. 3-13. Worn needles make accurate adjustment impossible.

pausing a few seconds between each adjustment. When the (fuel) rich limit is reached, engine rpm will falter and puffs of black smoke will issue from the exhaust. Two-cycle engines will "four stroke," that is, fire with a loud bang, miss on one revolution, and bang again.

5. Note position of screw at rich limit and, working in small increments, tighten the screw past the region of smooth running and to the point where rpm builds, hovers, and drops. This is *lean roll*, representing the least amount of fuel that will maintain combustion. Note the position of the screw.

6. Open the adjustment screw to the midpoint between lean roll and rich limit.

7. Close the throttle and adjust the low speed mixture control screw for fastest idle. It might be necessary to change the adjustment of the idle rpm screw.

8. Low-speed and high-speed adjustments are to some degree interdependent. Open the throttle as before and adjust the high-speed screw for best rpm. Drop the throttle back to idle and recheck the low speed adjustment.

9. With the engine idling, open the throttle quickly.

Caution: Utility engines—particularly those with splash lubrication—require a minimum 1600 rpm idle speed to assure oil circulation. This speed should be set with a tachometer.

10. Should the engine hesitate, open the high-speed needle 1/8 turn and retest until acceleration is rapid and smooth. This is an approximation of the true adjustment, even if the mixture is somewhat rich at high, no-load speeds.

11. The high-speed mixture can be fine-tuned by making final adjustments under load.

For single mixture control screw, typical of suction-lift carburetors, follow the first three steps, substitute the following three steps, and then continue.

4. With the throttle half open, adjust the single mixture screw for best rpm. Some slow rise and fall of engine speed may be apparent.

5. Close the throttle to a fast (1600 rpm) idle and, using your fingers, flip it open rapidly. If the engine stubles, back off the mixture screw 1/8 turn and repeat the acceleration test.

6. The correct adjustment may give an overly rich idle, but that is a design limitation, not easily corrected.

Throttle Slide Types. These carburetors theoretically have six adjustment points, each with its own domain across the range of throttle movement. However, parts availability and common sense reduce adjustments to three.

Remove the air filter and any plumbing that obstructs the view down the carburetor bore. Adjustment points are described as follows.

● Idle adjustment screw, present on all carburetors and bearing against the lower face of the throttle slide. The higher the slide is raised the faster the engine idles.

● Low speed mixture screw, present on all carburetors, located near the throttle slide and often regulates air (rather than fuel) flow. When this is the case, tightening the screw enriches the mixture between closed and 1/8 open throttle (A of Fig. 3-14).

● Cutaway, milled on the edge of the slide throttle to drive air into the bore. Increasing the angle of the cutaway leans the mixture between 1/8 and 1/4 throttle (B of Fig. 3-14). While all carburetors have such a cutaway, changing the factory adjustment depends upon parts availability and is rarely done.

● Fuel needle setting, or distance the needle extends below the throttle slide. Needle is usually grooved and secured to slide with a spring clip. Raising the needle, obscures less of the high speed jet and provides a richer mixture at 1/4 to 3/4 throttle (C of Fig. 3-14). Needles can sometimes be obtained with different tapers.

● High speed jet diameter controls mixture strength between 3/4 and wide open throttle (D of Fig. 3-14). Unless the engine has been extensively modified or is operated at high altitudes, a jet change is normally not required.

Troubleshooting

Although design differences always intrude, carburetors operate on the same general principles and the same broad troubleshooting procedures apply. Combustion chamber flooding, shown by a wet spark plug tip, can develop when the engine does not start after persistent cranking. The cause of the difficulty is usually in the ignition system (although improper carburetor settings can contribute).

Chamber flooding will usually clear of its own accord by waiting 20 minutes or so for the surplus gasoline to evaporate. Cranking with the throttle and choke wide open might help restarting, but

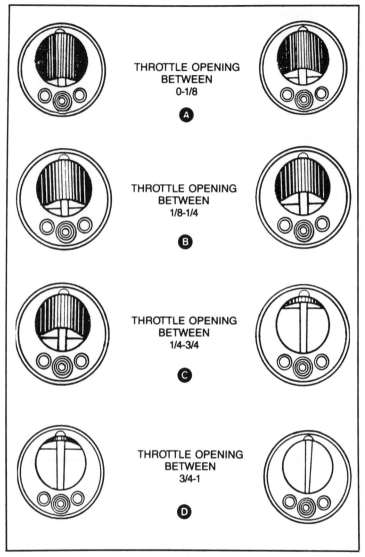

THROTTLE OPENING
BETWEEN
0-1/8

A

THROTTLE OPENING
BETWEEN
1/8-1/4

B

THROTTLE OPENING
BETWEEN
1/4-3/4

C

THROTTLE OPENING
BETWEEN
3/4-1

D

Fig. 3-14. Black, fluffy carbon deposits on a spark plug tip indicate an overly rich fuel-air mixture. (Courtesy of Kohler Co.)

the quickest fix is to change or dry the spark plug and boost magneto voltage. This is done by holding the spark plug wire 3/8 of an inch from the spark plug terminal and cranking. The air gap thus created wrings all available voltage from the ignition system.

Carburetor flooding is entirely another matter and can occur as soon as gasoline is introduced into the instrument. Fuel dribbles through the float bowl vent and, depending upon design, from the carburetor throat. This condition is hazardous; until corrected, do not attempt to start the engine.

Carburetor flooding is most often caused by a failure of the needle and seat to make a leak-proof seal. Sometimes a bit of dirt is the culprit. Nevertheless, it is always good practice to replace both parts of the inlet valve. Float-type carburetors can also flood because of a punctured, binding, or maladjusted float.

Fuel Starvation. Zero fuel delivery is easy to diagnose on a cold engine because the spark plug tip remains resolutely dry after prolonged cranking. Heroic efforts may oil the tip, but the characteristic odor of gasoline will be absent. Hot-engine diagnosis is more difficult because the spark plug tip usually remains dry. If there is fuel delivery, however, it should be possible to flood the chamber by cranking with the choke closed.

Stoppages most often develop upstream of the carburetor where flow pressure and velocity are low. Inspect screens at the tank and carburetor fittings, fuel filters, and fuel pump. Pump failure is almost always a diaphragm failure, although mechanical, automotive-type fuel pumps employ suction and discharge-side check valves that seem especially vulnerable to dirt. Another point to remember is that an upstream fuel line air leak can rob the pump of its prime.

As far as the carburetor is concerned, stoppages tend to be associated with the internal fuel-level mechanism. The check ball in suction-lift carburetor pickup tubes might stick in the closed position, the float might hang closed and, of course, the diaphragm could go haywire.

Obstructions in internal circuits are fairly rare, but can occur if the engine is stored without draining the tank and carburetor. A massive air leak at the mounting flange gasket can also deny fuel to the engine, but this condition is quite obvious because a leak of this magnitude implies that the carburetor jitters on its mounts. The ultimate cause might be a bent crankshaft.

Refusal to Idle. Assuming that the problem originates in the carburetor and is not a symptom of the throttle, governor, or ignition timing maladjustment, the cause is partial fuel starvation. Check for an obstruction in the low speed circuit and for a vacuum leak downstream of the throttle.

Low-speed circuit obstructions can often be cleared by removing the low-speed mixture control needle and gently opening the low-speed jet with a broom straw. Do not use wire, a drill bit, or other hard object that can score the jet and change its flow calibration. Compressed air will work when the jet is not visible from the vantage point of the mixture screw or when the obstruction is upstream of the jet.

Intake vacuum leaks—that is, leaks on the engine side of the throttle plate—are almost always confined to the carburetor mounting flange. Check the flange bolt tightness and gasket condition. Some engines are fitted with vacuum-operated accessories that might develop leaks and increase idle rpm.

As mentioned earlier, Walbro carburetors employ a cross-drilled main fuel nozzle. If this nozzle is disturbed, cross-drilled holes no longer index with the idle circuit and no fuel will pass. Repair by replacing the original nozzle with the undercut version that provides fuel—no matter how assembled.

Rich Running. Symptons of this malady are:

- Black smoke in exhaust.
- Soot on spark plug tip that will extend to inside of exhaust pipe (Fig. 3-14).
- Lack of power.
- Four-stroking with two-cycle engines.

Three conditions can cause rich running. The first and most common is too high an internal fuel level in the carburetor. This may be caused by a maladjusted float level, a fuel-sodden float (whether of plastic or brass construction), a sticking float (usually caused by too much float drop), a leaking inlet needle and seat and, for Tillotson diaphragm carburetors, a worn or bent diaphragm lever.

The second cause is mixture control maladjustment. This is remedied as described in the "Adjustments" section.

The third possibility is an air obstruction upstream of the throttle. Test the air cleaner by removing it while the engine is running. Some rpm increase might be experienced on utility engines because of designed-in restrictions, but rpm should not increase more than 100 rpm or so. A pronounced speed increase means that the filter element needs cleaning or replacement. It is also possible for the choke to pull into engagement on some designs, particularly if choke attenuation is automatic.

Lean Running. This condition has these symptoms:

- High cylinder head temperatures, shown by whitish spark plug tip (Fig. 3-15).

Fig. 3-15. White insulator, or rip, means dangerously high combustion-chamber temperatures, most often caused by an overly lean fuel-air mixture. (Courtesy of Kohler Co.)

- Lean roll, or a slow falling and building of engine rpm.
- Lack of power.
- Carburetor spit-back as the throttle is opened suddenly.
- Engine might pick up rpm with choke partially closed.

Lean running resolves into insufficient fuel or excessive air. Insufficient fuel delivery can usually be corrected as described in the "Adjustments" section. If this does not help, check the float adjustment and condition of Tillotson diaphragm lever. Massive air leaks downstream of the throttle can also cause this difficulty. Check the carburetor mounting flange surface and, on two-cycle engines, the condition of the main bearing seals.

Cleaning and Repair

Lacquer thinner is an adequate solvent for routine carburetor cleaning. More severe cases respond to Gunk Carburetor Cleaner—available from auto parts stores in pint containers—and very corroded carburetors need something on the order of Bendix Econo-Cleane. While Gunk seems fairly benign, Bendix rapidly attacks plastic and rubber parts and, upon several days immersion, leaches zinc castings.

When powerful chemicals are used, the carburetor must be disassembled far enough to remove plastic and other soft parts. Do not, in the normal course of events, disturb throttle plates, welch (expansion) plugs, or pressed-in fuel inlet fittings. Grind screwdrivers to fit jet and inlet seat slots.

The process goes something like this:

1. Obtain a carburetor overhaul kit that should contain all necessary replacement parts and instructions. Some kits include float-level gauges.

2. Remove the carburetor from the engine. Note how the governor link and related springs are attached to the throttle lever; adjust screw positions from closed.

3. Clean external carburetor surfaces.

4. Work on a clean bench, laying out parts in order of disassembly. Refer to drawing in overhaul kit.

5. Immerse semi-stripped carburetor body, jets and other metallic parts in cleaner for at least 30 min.

6. Rinse cleaner from parts as instructed on label.

7. Assemble unit, using kit parts.

Needle and Seat. A few carburetors still use metallic needle and seat assembles, but the tendency now is to substitute plastic for one or both of these parts. Figure 3-16 shows recent and current Walbro types, both with elastometer-tipped needles with brass seats. Forcing the needle into the seat, as when making float adjustments, could permanently deform the needle tip and result in a fuel leak.

Figure 3-17 depicts another Walbro type that features a Viton seat and plastic needle. The old seat is extracted with a hooked wire and the new seat is pressed into place with its ringed side toward the carburetor body. Lightly oiling the seat cavity aids installation. Both the current brass seat design (shown on the right of Fig. 3-16) and the Viton seat-type employ a float damper spring. Figure 3-18 describes spring installation.

Briggs & Stratton elastomer seats, used on Flo-Jet carburetors, are removed with a self-tapping screw and installed using the original seat as a ram. See Fig. 3-19. This seat must be flush with its boss.

Throttle Shaft Bushings. Industrial-type engines employ bushings at the throttle shaft that should be replaced, together with

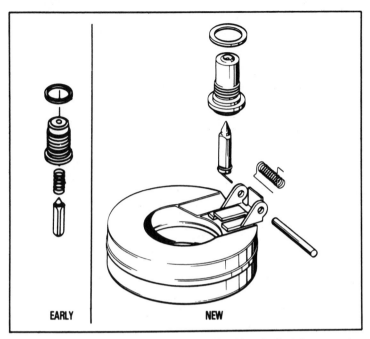

Fig. 3-16. Two Walbro needle and seat assembles. Note the float damper spring shown on the right and discussed in captions with Figs. 3-18 and 3-19.

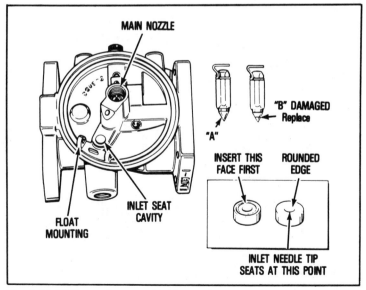

Fig. 3-17. Walbro carburetor with Viton seat.

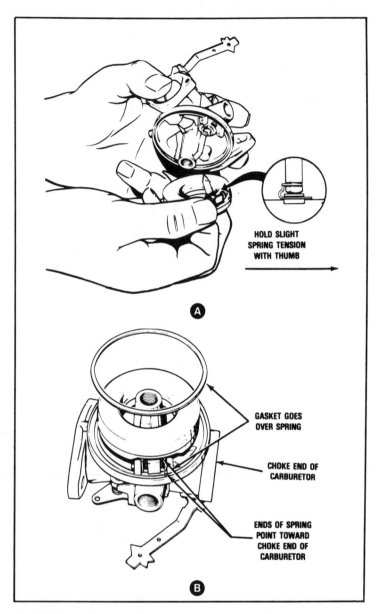

HOLD SLIGHT
SPRING TENSION
WITH THUMB

A

GASKET GOES
OVER SPRING

CHOKE END OF
CARBURETOR

ENDS OF SPRING
POINT TOWARD
CHOKE END OF
CARBURETOR

B

Fig. 3-18. Assemble current Walbro carburetors with needle (hair)spring hanging down and ends of float damper (coil) spring pointed at choke. Carefully place coil spring between float hinges and assemble needle to float tab. Now wind the coil spring back to put tension on it and install the float over the hinges, securing the assembly with the hinge pin (A). Install the bowl gasket over the hinge pin (B).

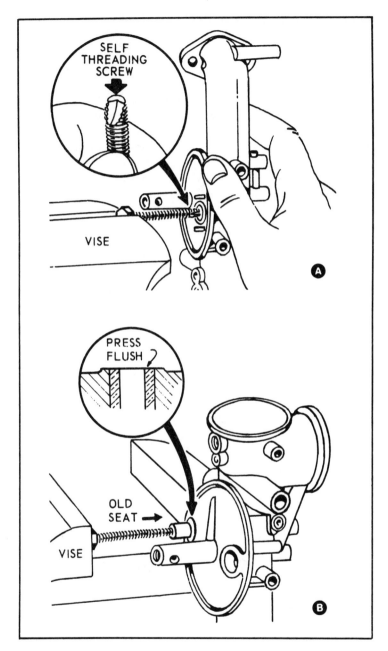

Fig. 3-19. Elastomer seat in Briggs & Stratton Flo-Jet can be removed with aid of self-tapping screw and vise (A). Install new seat flush with recess, using original seat as ram (B).

the shaft, when side play exceeds 0.008-0.010 of an inch (Fig. 3-20). The throttle plate may not be reversible and the choke side should be scribe marked before detaching the plate from the throttle shaft. (Some Walbro throttle plates have a reference mark at 12 and/or 3 o'clock.) Remove the old bushings with an appropriately sized tap or Ease-Out—usually 1/4 inch—and press in new bushings. Ideally, bushings should be reamed, but this is not considered absolutely necessary. Install a new throttle shaft and lightly tighten the hold-down screw(s). Shut the throttle, centering the plate in the carburetor bore. Tighten the screws, staking them if that is factory practice. Test the throttle for possible binds.

Welch Plugs. Normally it is not a good idea to disturb welch, or expansion, plugs used to seal large cavities in carburetor bodies. But these plugs may develop leaks and sometimes must be removed to access primer pump valves and other vulnerable parts. First, make certain that you can obtain a replacement plug and then, using a small capehart chisel, pierce the plug (Fig. 3-21A). Press down on the chisel, prying the plug out of its recess. Install a new plug—convex side out—with a punch that has the same or slightly larger diameter as the plug (B of Fig. 3-21). Seat the new plug with its edges just shy of the carburetor casting surface. Avoid flattening the plug because spring tension developed across plug convexity maintains the seal. No chemical sealant is necessary or desirable.

Primer Pump. Some diaphragm and float-type carburetors employ a primer pump (rather than a choke valve) as a cold starting aid. The pump bulb might be located at some distance from

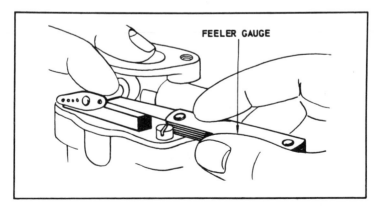

Fig. 3-20. Throttle bushings are subject to fairly severe wear and should be renewed when throttle shaft to bushing clearance is more than 0.008 of an inch or so.

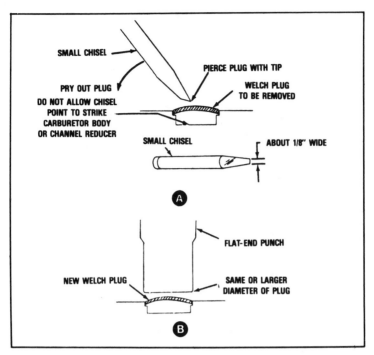

SMALL CHISEL

PIERCE PLUG WITH TIP

PRY OUT PLUG

WELCH PLUG
TO BE REMOVED

DO NOT ALLOW CHISEL
POINT TO STRIKE
CARBURETOR BODY
OR CHANNEL REDUCER

SMALL CHISEL

ABOUT 1/8" WIDE

Ⓐ

FLAT-END PUNCH

NEW WELCH PLUG

SAME OR LARGER
DIAMETER OF PLUG

Ⓑ

Fig. 3-21. Welch plugs are removed by first puncturing the plug with a chisel, then levering it up and out (A). New plugs are seated to the depth shown with a flat punch (B).

the carburetor or directly below it as an integral part of the fuel bowl hold-down bolt. In either case, access is easy and obvious. However, some designs incorporate the pump bulb on the side of the main body casting, adjacent to the high speed fuel circuit. Remove the pump bulb and retainer with main force (as shown in Fig. 3-22).

Castings. Mounting flanges are usually secured to the cylinder by two capscrews. Because the gasket is thick and resilient, overtightening the screws invariably distorts the casting. The gasket surface can be restored with an Armstrong "surface mill." Tape a sheet of medium-grit emory cloth to a piece of plate glass or a drillpress work table and, using a circular motion, grind until the gasket surface takes a uniform sheen.

Up-and-downdraft carburetor float bowl covers are also vulnerable to distortion. Check cover-to-bowl clearance as shown in Fig. 3-23 and correct by gently tapping casting "ears" with a very small hammer.

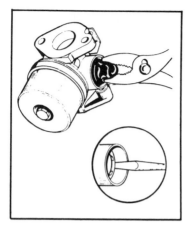

Fig. 3-22. In this configuration, the primer bulb and retainer washer are disposable items, removed as shown. Install replacement parts using a 3/4-inch deep well socket as a pilot.

AIR FILTERS

With the exception of outboard motors, most small engines are equipped with an air filter. Two-cycle engines may employ a wire mesh or composite fiber filter, which should routinely be cleaned with kerosene and re-oiled. The "fuel fog" that hovers in front of two-cycle carburetors is intended to keep the filter element wet and working between service intervals.

A design weakness of most of these filters is the single-screw

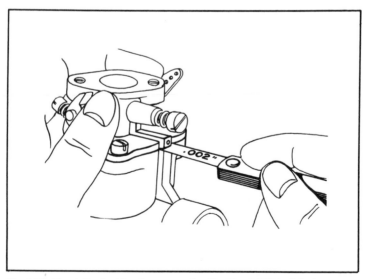

Fig. 3-23. Check the float bowl cover casting fit with a 0.002-inch feeler gauge. If the gauge will enter, remove the casting and straighten with light hammer taps.

mounting arrangement that vibrates loose in service, allowing grit to enter the engine. Correct by securing the screw with red Locktite or equivalent thread adhesive.

Industrial engines often employ centrifugal oil bath filters (Fig. 3-24). Air entering the top of the unit makes an S-turn that centrifuges out heavy solids. The mesh, continuously wetted by oil in the reservoir, traps smaller particles. These filters should be cleaned in solvent and replenished with straight—not multigrade—30-weight motor oil. Overfilling or using a lighter oil can cause the engine to smoke.

Pleated paper filters are among the most efficient in terms of particle size trapped and permeability to air, but these filters cannot, in any real sense, be cleaned and must be periodically replaced. (Liquids swell the paper fibers, blocking air entry.)

Polyurethane, or sponge, filters (Fig. 3-25) match pleated paper efficiency with the added bonus of reusability. Wash in solvent or water and deterrent and re-oil with a few cc's of motor lube, knealing the element to distribute the oil. It is good practice to check the element each time the engine is started.

Some manufacturers employ a polyurethane "precleaner" in conjunction with a pleated paper main element (Fig. 3-26). The

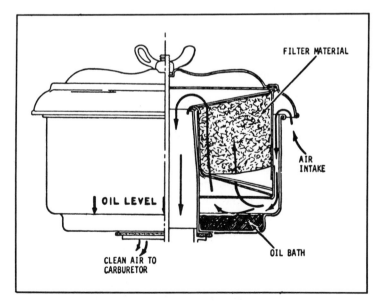

Fig. 3-24. Oil-bath filters do a fair job and are at least reusable. This is an early Kohler type.

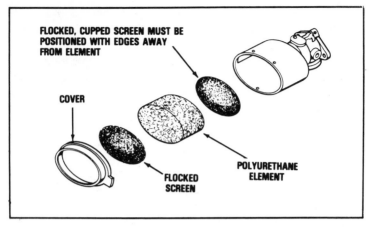

Fig. 3-25. Polyurethane filters achieve good results when properly wetted, but suffer from oil migration and need frequent attention to maintain full efficiency.

precleaner has little depth of filtration and requires frequent attention. Clean with detergent and water, allowing it to air dry before oiling. Gently squeeze—do not wring—excessive oil from the sponge that would otherwise migrate to the paper element.

GOVERNORS

Industrial and utility engines incorporate a governor to help maintain engine rpm under varying loads and, more importantly, to limit maximum speed to the friendly side of 3,600 rpm. Outboard, chain saw and other high-performance engines do not, in general, employ such a governor, although many have some form of rpm limiter (usually tied into the ignition system).

Air vane governors, such as the own shown in Fig. 3-27, sense engine speed as a function of cooling air pressure and velocity. The air vane is installed under the shroud and in the cooling stream. It is spring-loaded and connected to the throttle by a linkage. As engine rpm increases, air pressure on the vane reacts against spring tension to shut the throttle, slowing the engine. If the engine speed decreases past a certain limit, the spring opens the throttle.

Figure 3-26 shows a fixed-speed governor. Engine speed is fixed in the sense that the operator has no throttle and can only adjust the speed by loosening a screw and moving the spring-anchor bracket. A variable-speed governor, one that gives the operator discretion over engine speed, is similar in construction, but employs a moveable spring anchor. Opening the throttle control stretches the spring, increasing tension on it and causing the carburetor throttle plate to open wider. Closing the throttle control relaxes spring tension and biases the system in favor of the air vane that acts to close the carburetor throttle plate.

Centrifugal governors sense engine rpm as flyweight movement (Fig. 3-27). Camshaft-driven flyweights respond to increasing engine rpm by moving outward. This movement, acting through a spool

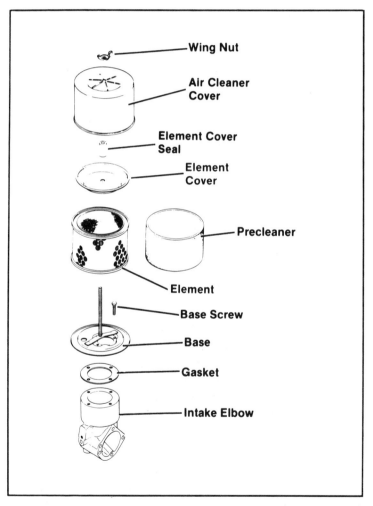

Fig. 3-26. Modern Kohler filter combines paper element with polyurethane precleaner.

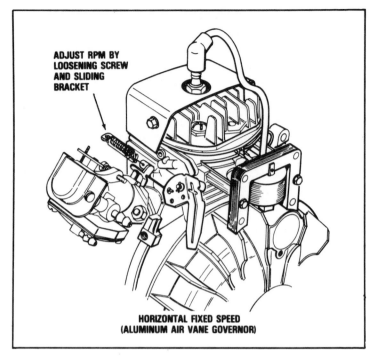

Fig. 3-27. Engine speed is changed on this fixed-speed air vane governor by loosening screw and moving balance spring anchor bracket. The design is used on some Tecumseh two-cycle engines.

and yoke mechanism, rotates the governor shaft and attached governor lever. The lever is linked to the carburetor throttle plate and tends to close the throttle as rpm increases. A small coil spring, known as an *angleich* or balance spring, opposes flyweight action and would keep the throttle full open.

When the engine is shut down, flyweights are neutralized and the balance spring is free to pull the throttle plate open. As soon as the engine starts, flyweights energize to close the throttle to the extent balance-spring tension permits. Under load, engine speed drops, flyweights retract and the balance spring opens the throttle. Engine speed picks up, flyweights respond and a new equilibrium is established.

Fixed-speed governors have their balance springs more or less permanently anchored so that spring tension remains constant. On variable-speed designs, the spring anchor is connected to the hand throttle and tension can be varied at will over the rpm range. This arrangement is clearly shown in B of Fig. 3-28.

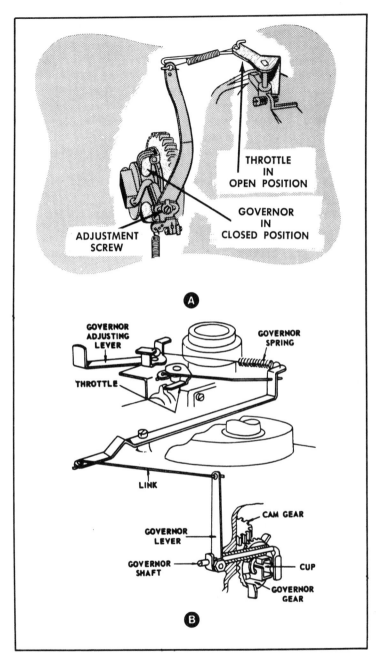

THROTTLE
IN
OPEN POSITION

GOVERNOR
IN
CLOSED POSITION

ADJUSTMENT
SCREW

A

GOVERNOR
ADJUSTING
LEVER

GOVERNOR
SPRING

THROTTLE

LINK

CAM GEAR

GOVERNOR
LEVER

GOVERNOR
SHAFT

CUP

GOVERNOR
GEAR

B

Fig. 3-28. Centrifugal governors as applied to horizontal (A) and vertical crankshaft (B) engines.

107

Some centrifugal governors can be adjusted for sensitivity, usually, by changing the mechanical advantage of the balance spring relative to the governor lever (Fig. 3-29). Insufficient sensitivity allows the engine to lose excessive speed under abrupt loads; too much sensitivity, on the other hand, can result in hunting or continuous speed corrections.

Other adjustments are possible. Some designs incorporate a low-speed adjustment (distinct from the carburetor idle-speed screw) and all centrifugal governors allow changes in the factory high speed limit. A few governors incorporate a screw for this purpose. Others can be fiddled by changing the position of the balance spring relative to the throttle or by bending the linkage.

Warning: Use an accurate tachometer when making low speed and—most emphatically—high-speed adjustments. Do not exceed factory rev limits.

The governor can be recalibrated. This operation is necessary when the unit has been dismantled or severely worn (Fig. 3-30). The theory of this adjustment is the same for all governors: with the engine shut down, loosen the governor lever on the governor shaft and move the lever to the full-open throttle position. Holding the lever with one hand, turn the governor shaft to seat the yoke against the flyweights. The problem with this adjustment is that the direction of governor shaft movement varies with engine manufacturer and model. A mistake will dramatically destroy the governor and usually takes out part of the cylinder block as well. Therefore, obtain factory instructions or the help of a mechanic familiar with the engine model in question *before* attempting this adjustment.

FUEL PUMPS

Small engine fuel pumps are either mechanical or diaphragm types. Either can be tested as follows:

1. Determine that fuel reaches the pump. If fitted, remove the inlet filter to assure fuel passage.
2. Crack the outlet line fitting at the pump or carburetor.
3. Switch off the ignition.
4. Crank the engine.
5. If the fitting remains dry, the pump does not function and must be repaired.

Other signs of pump failure include a fuel starvation under full

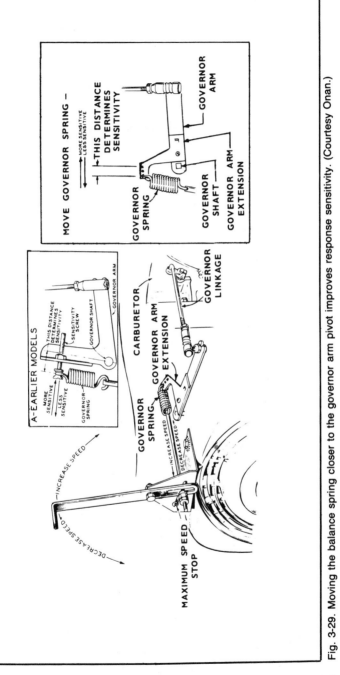

MOVE GOVERNOR SPRING –

MORE SENSITIVE
LESS SENSITIVE

THIS DISTANCE DETERMINES SENSITIVITY

GOVERNOR ARM

GOVERNOR SPRING

GOVERNOR SHAFT

GOVERNOR ARM EXTENSION

A-EARLIER MODELS

MORE SENSITIVE
LESS SENSITIVE

THIS DISTANCE DETERMINES SENSITIVITY

SENSITIVITY SCREW

GOVERNOR SHAFT

GOVERNOR ARM

GOVERNOR SPRING

CARBURETOR

GOVERNOR ARM EXTENSION

GOVERNOR LINKAGE

GOVERNOR SPRING, GOVERNOR ARM EXTENSION

INCREASE SPEED

DECREASE SPEED

INCREASE SPEED

DECREASE SPEED

MAXIMUM SPEED STOP

Fig. 3-29. Moving the balance spring closer to the governor arm pivot improves response sensitivity. (Courtesy Onan.)

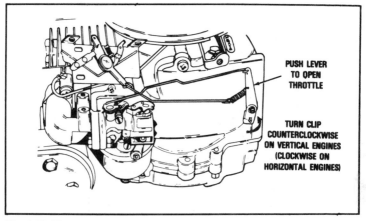

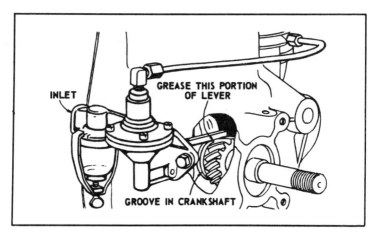

Fig. 3-30. The relationship between governor shaft and wide open throttle is probably the crucial aspect of small engine service work. Unfortunately, procedures are not standardized (as this Tecumseh illustration shows).

load and throttle, and gasoline (detected by its odor) in the crankcase.

MECHANICAL PUMPS

Mechanical pumps are, for the most part, miniaturized automotive units that drive from a camshaft or crankshaft eccentric. Figure 3-31 illustrates a Briggs & Stratton pump used on several

Fig. 3-31. Briggs & Stratton mechanical fuel pump drives from a crankshaft eccentric.

of the company's larger engines. Replace the diaphragm with the following procedure:

1. Remove the pump assembly from the engine.

2. Separate pump head (or upper body casting) from pump body by removing the four hold-down screws. Note position of fuel outlet fitting relative to the fuel line.

3. Using a small punch, drive out the pin in either direction far enough to disengage pump lever.

4. Unhook lever from diaphragm. Remove lever, lever spring and diaphragm.

5. Position replacement diaphragm over diaphragm spring with slot in diaphragm shaft turned 90° from lever.

6. Install lever without lever spring (Fig.3-32A). Hook end of lever into diaphragm shaft slot.

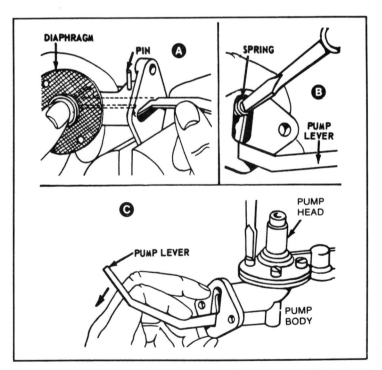

Fig. 3-32. Service procedures for the Briggs & Stratton pump: Displace pin far enough to release lever (A), install lever spring after lever and diaphragm engagement is made (B), and compress diaphragm spring before tightening pump head screws (C).

7. Insert lever spring into position above lever. Inner end of spring fits over projection in pump body casting; outer end is jimmied over lever extension with a screwdriver (Fig. 3-32B).

8. Start—but do not run down—screws holding pump body head to body casting. See that pump head fitting aligns with fuel pipe.

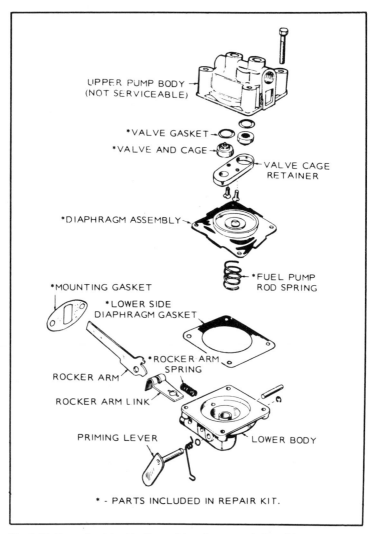

Fig. 3-33. Pump furnished by Onan drives from camshaft and features a priming lever.

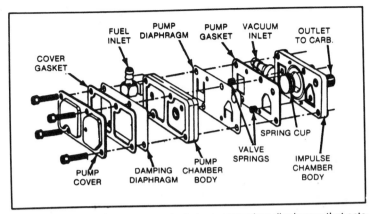

Fig. 3-34. Briggs & Stratton pump includes a secondary diaphragm that acts like an accumulator to dampen fuel pulses. Check valves are integral with main diaphragm. Tecumseh employs a similar design.

9. Press down hard on lever and tighten head screws in an X-pattern. Steps 8 and 9 prevent excessive diaphragm stretch.

10. Position a new fuel pump flange gasket on the engine block, securing it with Permatex or an equivalent gasket adhesive.

11. Mount the pump, making certain that the pump lever rides on the flanged crankshaft eccentric. Start the pump-to-block capscrews.

12. Tighten screws, turning flywheel as necessary to relieve tension on the pump lever.

13. Make up fuel line connections.

14. Start engine to verify that pump operates and that there are no fuel leaks.

Figure 3-33 illustrates a somewhat more elaborate mechanical pump used by Onan and Kohler. Service procedures generally parallel those for the Briggs & Stratton unit and, at the risk of repetition, the pump diaphragm spring must be compressed before upper body screws are torqued down. Otherwise the diaphragm will be overstressed and fail rapidly.

Diaphragm

Diaphragm pumps, also known as impulse, pulse, vacuum, and pressure differential pumps, rely on the changing crankcase vacuum to generate diaphragm movement (Figs. 3-34 and 3-35). As the piston moves toward the top of its stroke, a partial vacuum is created

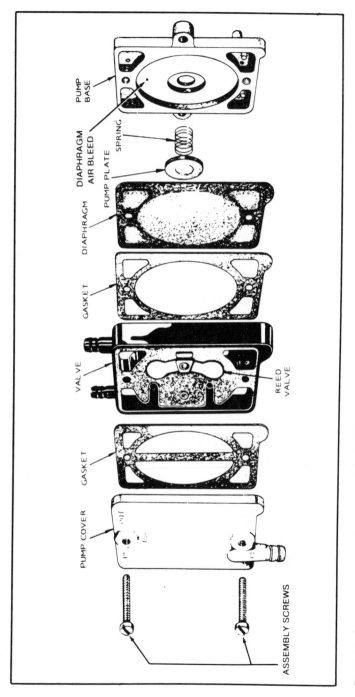

Fig. 3-35. Onan pump is a serious piece of work (with reed valves).

Labels in figure:
PUMP BASE
DIAPHRAGM AIR BLEED
SPRING
PUMP PLATE
DIAPHRAGM
GASKET
VALVE
REED VALVE
GASKET
PUMP COVER
ASSEMBLY SCREWS

in the crankcase. This vacuum, transmitted to the pump by an impulse line or, on block-mounted units, by a port, displaces the diaphragm to allow fuel to enter the pump cavity. Crankcase vacuum almost entirely dissipates on the downstroke and the diaphragm, impelled by its spring, moves to force fuel through the pump outlet.

Nearly all pump failures can be corrected by replacing the diaphragm. However, a stoppage at the atmospheric bleed port or leaks in the inlet or impulse plumbing can also disable the pump. Some designs include an air bleed filter that requires periodic cleaning and re-oiling.

To service the pump, disconnect fuel and vacuum lines, and remove the unit from its mounting bracket. Scribe a mark across the pump stack as an assembly aid. Carefully remove the parts, strata by strata, noting their sequence and orientation. Pay particular attention to the position of the diaphragm gasket relative to the diaphragm. Some are on the outboard side of the diaphragm and some are under it. Clean metallic parts in solvent and replace soft parts from the overhaul kit. Install assembled pump on engine and check for leaks.

Chapter 4

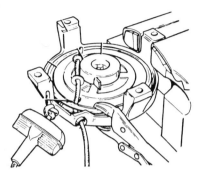

Rewind Starters

Unlike other engine systems that operate continuously, manual and electric starters are designed for intermittent use. This is why rewind starters can get by with nylon bushings and motor pinions can cheerfully bang into engagement with the flywheel. The starter usually lasts about as long as the engine and the owner is satisfied.

But the balance between starter and engine life goes awry if the engine is allowed to remain chronically out of tune. Most starter failures are the result of overuse: the starter literally works itself to death cranking a baulky engine. The mechanic must repair the starter and—if the repair is to be permanent—must correct whatever it is that makes the engine reluctant to start.

SIDE PULL

The side-pull rewind (AKA recoil, self-winding, and retractable) starter was introduced by Jacobsen in 1928 and has changed little in the interim. These basic components are always present (Fig. 4-1).

- Pressed steel or aluminum housing, which contains the starter and positions it relative to the flywheel.
- Recoil spring, one end of which is anchored to the housing, the other to the sheave.
- Starter rope (nylon, although a few Fairbanks Morse

wirelines are still encountered) which is anchored to and wound around the sheave.

- Sheave or pulley.
- Sheave bushing between sheave and housing or (on Briggs & Stratton) between sheave and crankshaft.
- Clutch assembly.

Troubleshooting

Most failures have painfully obvious causes, but it might be useful to have an idea of what you are getting into before the unit is disassembled.

- *Broken rope*—the most common failure, often the result of excessive tension on the rope near the end of its stroke or by pulling the rope at an angle to the housing. The problem is exacerbated by a worn rope bushing (the guide tube, at the point where the rope exits the housing). In general, rope replacement means complete starter disassembly, although some designs allow replacement with the sheave still assembled to the housing.
- *Loss of spring tension*—usually the result of a broken spring, but may also be caused by spring disengagement from the housing or sheave. The spring anchor slot on Briggs & Stratton housings may batter out and release the spring. Complete disassembly is required.
- *Rope fails to completely retract*—on a unit in service, suspect loss of spring preload tension due to aging. Best recourse is to replace the spring, although preload tension can be increased by one sheave revolution. On a just-repaired unit, check the starter housing/flywheel alignment, spring preload tension, and replacement rope length and diameter.
- *Failure to engage flywheel*—a clutch problem, caused by a worn or distorted brake spring (Fig. 4-1 illustrates a coil-type brake spring; other designs employ more vulnerable hair springs), retainer screw backout, or oil on clutch friction surfaces. While recoil spring and sheave bushing generally require some lubrication, starter clutch mechanisms must, as a rule, be assembled dry.
- *Excessive force required to pull rope*—check starter housing/flywheel alignment first. Then remove starter, turn the engine over by hand to verify that it is free, and check starter action. Problem might involve a dry sheave bushing.
- *Noise from starter as engine runs*—check starter housing/fly-

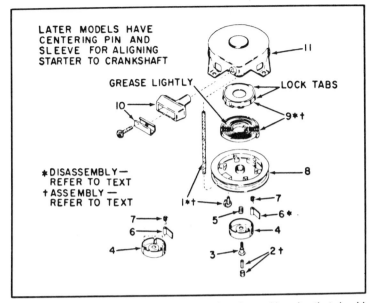

Fig. 4-1. Eaton rewind starter with integral mainspring and housing that should not be dismantled in the field. These starters can be recognized by lock tabs on the spring housing OD. This starter also uses a small coil spring—shown directly below the sheave—to generate friction on the clutch assembly.

wheel alignment. On Briggs & Stratton designs, the problem is often caused by a dry sheave bushing (located between the starter clutch and crankshaft). Remove the blower housing and apply a few drops of oil to the crankshaft end.

SERVICE PROCEDURES

Rewind starters are a special technology, and it is helpful to take an overall view of the subject. The first order of business is to release spring preload tension. There are two ways to do this. Any rewind starter can be disarmed by removing the rope handle and allowing the sheave to unwind in a controlled fashion. Other starters have provision for tension release with the handle still attached to the rope. Briggs & Stratton provides clearance between sheave OD and housing that allows several inches of rope to be fished out of the sheave groove. This increases the effective length of the rope, enabling sheave and attached spring to unwind. Many other designs incorporate a notch in the sheave for the same purpose.

Brake the sheave with your thumbs as it unwinds. It is also

119

good practice to number sheave rotations from the point of full rope retraction so that the same preload can be applied on assembly.

The sheave is secured at its edges by crimped tabs and located by the crankshaft extension (Briggs & Stratton side pull), or else rotates on a pin attached to the starter housing. A screw (Eaton) or retainer ring (Fairbanks-Morse, several foreign makes) secures the sheave to the post.

The main spring lives under the sheave, coiled between sheave and housing, with its inner, or movable, end secured to the sheave hub. The outer, or stationary, spring end anchors to the housing.

Warning: Even after preload tension is dissipated, rewind springs store energy that can erupt when the sheave is disengaged from the housing. Wear safety glasses.

The manner in which recoil springs secure to the housing varies among makes, and this affects service procedures. Some Eaton starters use an integral spring retainer that indexes to slots in the housing (Fig. 4-2). Spring and retainer are handled as a unit and should not be disassembled.

What might be called the standard attachment strategy is to secure the spring to a post, pressed into the underside of the housing. The fixed end of the spring forms an eyelet or, as the case may be, a hook, that slips over the anchor post. To simplify assembly, most manufacturers supply replacement springs coiled in a retainer clip. The mechanic positions the spring and retainer in the housing cavity with the spring eyelet over the post and

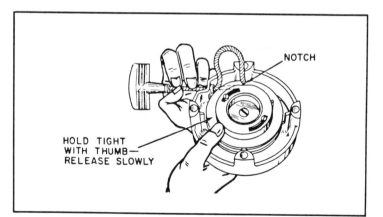

Fig. 4-2. Common sense dictates that the starter should be disarmed before sheave is detached. Most have provision to unwind the rope a turn or so while others are disarmed by removing the rope handle and allowing the rope to fully retract.

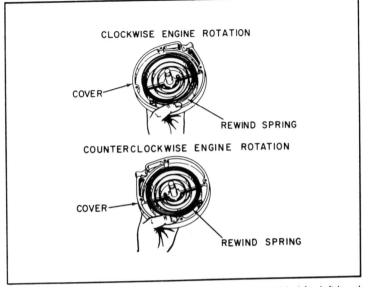

CLOCKWISE ENGINE ROTATION

COVER

REWIND SPRING

COUNTERCLOCKWISE ENGINE ROTATION

COVER

REWIND SPRING

Fig. 4-3. Many rewind springs and all ropes can be assembled for left-hand or right-hand engine rotation. This feature is a manufacturing convenience that makes life difficult for mechanics.

presses the spring out of the retainer (which is then discarded). Sheave engagement usually takes care of itself. Exceptions are discussed in sections dealing with specific starters.

Some starters adapt to left or right-hand rotation by reversing the spring (Fig. 4-3). Viewing the starter housing from the underside and using the movable spring end as reference, clockwise engine rotation demands counterclockwise spring wind up. The wrap of the rope must, of course, provide appropriate sheave rotation.

The third type of spring anchor—after the integral retainer and the housing post. This anchor takes the form of a slot in the starter housing through which the spring passes. Figure 4-4 shows a Briggs & Stratton side-pull unit that is similar to several OMC Lawnboy types.

These starters are assembled by winding the spring home with the sheave. Thread the movable end of the spring through the housing slot, engage the movable end with the sheave, and rotate the sheave opposite engine rotation until the whole length of the spring snakes through the housing slot. The fixed end of the spring is notched or hooked for retention by the slot.

Rewind spring preload is necessary to maintain some rope tension when the rope is retracted. Too little preload and the rope handle droops; too much and the spring binds solid to pull out of its anchors.

Depending upon starter make and model, either of two approaches is used to establish preload. Most manufacturers suggest this general procedure:

1. Remove the rope handle if it is still attached.
2. Secure one end of the rope to the sheave anchor.
3. Wind the rope completely over the sheave, so that the sheave will rotate in the direction of engine rotation when the rope is pulled.
4. Wind the sheave against engine rotation, a specified number of turns. If the specification is unknown, wind until the spring coil binds and release sheave for one or two revolutions.
5. Without allowing the sheave to unwind further, thread the rope through the guide tube (also called a ferrule or eyelet) in the starter housing and attach the handle.
6. Gently pull the starter through to make certain that the rope extends to its full length before the onset of coil bind and that the rope retracts smartly.

Another technique can be used when the rope anchors to the inboard (engine) side of the sheave:

1. Assemble sheave and spring.
2. Rotate the sheave, winding the mainspring until coil bind occurs.

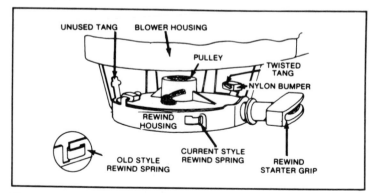

Fig. 4-4. Briggs & Stratton rewind starter used on 6 through 11 cubic inch engines. A later variant employs redesigned sheave and dispenses with nylon bumpers.

3. Release spring tension by one or no more than two sheave revolutions.

4. Block the sheave to hold spring tension. Some designs have provision for a nail that is inserted to lock the sheave to the housing; others can be snubbed with Vise-Grips or C-clamps.

5. With rope handle attached, thread rope through housing ferrule and anchor it to the sheave.

6. Release the sheave block and, using your thumbs for a brake, allow the sheave to rewind, pulling rope after it.

7. Test starter operation.

The starter rope should be the same weave, diameter, and length as the originals. If required length is unknown, anchor the rope to the sheave, wind the sheave until coil bind—an operation that also winds the rope on sheave—release the sheave for one or two turns, and cut the rope (leaving enough surplus for handle attachment).

Three types of *clutch assemblies* are encountered: Briggs & Stratton sprag, or rachet, Fairbanks-Morse friction-type, and the positive-engagement dog-type used by other manufacturers. In event of slippage, clean the Briggs clutch and replace the brake springs on the other types. Fairbanks-Morse clutch shoes may respond to sharpening.

One last general observation concerns *starter positioning*. Whenever a rewind starter has been removed from the engine or has vibrated loose, starter clutch/flywheel hub alignment must be re-established. Follow this procedure:

1. Attach starter or starter/blower housing assembly loosely to engine.

2. Pull starter handle out about 8 inches to engage the clutch.

3. Without releasing the handle, tighten the starter hold-down screws.

4. Cycle starter a few times to check for possible clutch drag or rope bind. Reposition as necessary.

BRIGGS & STRATTON

Briggs & Stratton side-pull starters are special in several respects (Fig. 4-4). In addition to its basic function of transmitting torque from the starter sheave to the flywheel and disengaging when the engine catches, the starter clutch also serves as flywheel nut and starter sheave shaft. Starter and blower housing assemblies are integral. It is possible, however, to drill out the spot welds and replace the starter assembly as a separate unit. Bend-over tabs locate the starter sheave in the starter housing.

Disassembly

Follow this procedure:

1. Remove blower housing and starter from engine.
2. Remove rope by cutting the knot at the starter sheave— visible from underside of blower housing.
3. Using pliers, grasp the protruding end of the mainspring and pull out as far as possible (Fig. 4-5). Disengage the spring from the sheave by rotating the spring a quarter turn or by prying one of the tangs up and twisting the sheave.
4. Clean and inspect. Replace the rope if it is oil soaked or frayed. Although it might appear possible to reform the end of a broken B & S mainspring, such efforts are in vain and the spring will have to be replaced for a permanent repair. The same holds for the spring anchor slot in the housing. Once an anchor has swallowed a spring, the housing should be renewed.

Assembly

1. Dab a spot of grease on underside of steel sheave. Note that a plastic version requires no lubrication. See Fig. 4-6.
2. Secure blower housing—engine side up—to the workbench.

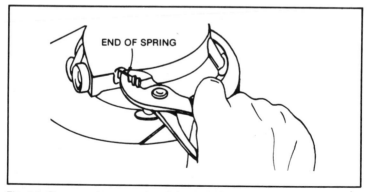

Fig. 4-5. Once rope is removed, pull the rewind spring out of the starter housing. The spring can be detached from sheave by twisting either sheave or spring a quarter turn.

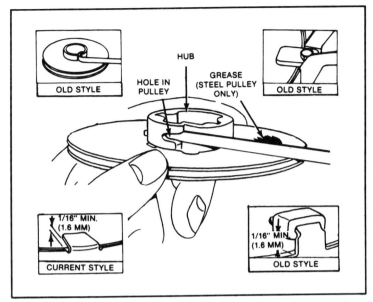

Fig. 4-6. Spring installation varies slightly with the date of manufacture. Steel sheaves require lubrication.

This can be accomplished with nails or C-clamps.

3. Working from outside of the blower housing, pass the inner end of mainspring through housing anchor slot. Engage inner end with the sheave hub.

4. Some mechanics attach rope—less handle—to the sheave at this point. The rope end is cauterized in an open flame and is knotted.

5. Bend tabs to give the sheave 1/16 of an inch endplay. Use nylon bushings on early models so equipped.

6. Using a 3/4-inch wrench extension bar or a piece of one by one inserted into sheave centerhole, wind the sheave 16 turns or so counterclockwise until full length of mainspring passes through the housing slot and coil binds.

7 Release enough mainspring tension to align the rope anchor hole in the sheave with the housing eyelet.

8. Temporarily block the sheave to hold spring tension. This can be done with a Crescent wrench snubbed between winding tool and blower housing. See A of Fig. 4-7.

9. If rope has been installed, extract end from between sheave flanges, thread through eyelet, cauterize and attach handle. If the rope has not been installed, pass the cauterized end through the

eyelet from outside the housing, between sheave flanges and out through sheave anchor hole (Fig. 4-7). Knot the end of the rope. Old-style sheaves incorporate a guide lug between flanges. The rope must pass between the lug and sheave hub. This operation is aided by a small screwdriver or a length of piano wire (A of Fig. 4-7).

The clutch is normally not opened unless it slips, a condition that may be caused by wear or presence of lubricant. Old-style assemblies are secured with a wire retainer clip; newer models depend upon retainer-cover tension and can be pried apart with a small screwdriver (Fig. 4-8). Clean parts with a dry rag (avoiding use of solvent). The clutch housing can be removed from the crankshaft using a special factory wrench described in Chapter 3.

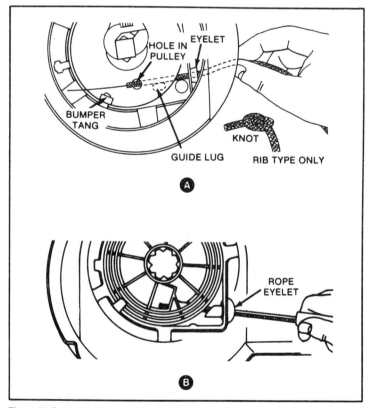

Fig. 4-7. Cast-iron block Briggs & Stratton engines never seem to wear out and starters with internal rope guide lugs are still encountered. Use a length of piano wire to push the top past the inner side of the lug as shown (A). Newer designs omit the guide lug, making installation easier (B).

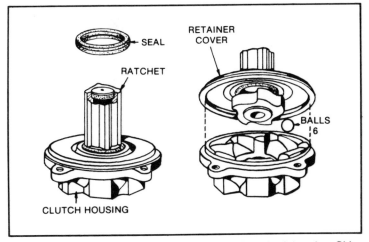

Fig. 4-8. Current production clutch cover is a snap fit to clutch housing. Older version employed a spring wire retainer. As a point of interest, older engines can be modified to accept new clutch assembly by trimming 3/8 inch from the crankshaft stub and 1/2 inch from sheave hub.

Eaton

Recognized by P-shaped engagement dogs, or pawls, Eaton starters are found on a wide variety of American engines. Light duty models employ a single pawl, more substantial types use two, and high-torque models have three. Eaton pioneered the use of mainspring and spring retainer (a feature that makes life easier and, perhaps, safer for mechanics). Another Eaton feature is the centering pin; it usually rides on a nylon bushing. Figure 4-1 shows a single-pawl Eaton starter with a mainspring and centering pin.

The most common complaint is failure to engage the flywheel. This difficulty can be traced to the clutch brake, which generates friction that translates into pawl engagement, or to the pawls themselves. Two brake mechanisms are encountered. The latest arrangement, shown in Fig. 4-1, employs a small coil spring that reacts against the cuplike pawl retainer. You can see the spring directly below the sheave and to the left of the pawl. Another brake mechanism, used for many years and apparently still in production, interposes a star-shaped brake washer between the pawl retainer and brake spring. Figures 4-9 and 4-12 illustrate this part. A shouldered retainer screw secures the assembly to the sheave and preloads the brake spring (Fig. 4-10).

Check the retainer screw, which should be just short of hernia tight, inspect friction parts, with special attention to optional star

127

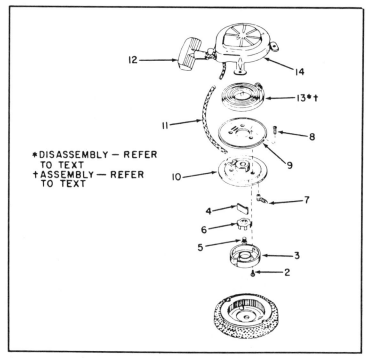

Fig. 4-9. Eaton light-duty pattern found on small two and four-stroke engines. This starter is distinguished by its uncased mainspring (13) and single-dog clutch (dog shown at 4, clutch retainer at 3). In event of slippage during cranking, replace friction spring 5 and brake 6.

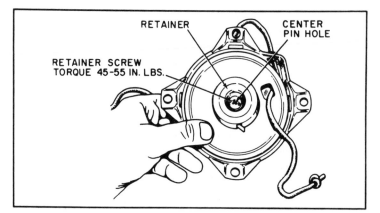

Fig. 4-10. View looking toward the inboard side of the sheave with one of two dogs assembled over return spring and brake spring in place. This unit is to be assembled dry; only snow-proof models, distinguished by half-moon cam that engages the dogs, require oil on dog-mounting posts.

brake, and check the pawl return spring (Fig. 4-11) that can be damaged by engine backfire. Clean parts, assemble without lubricant, and observe pawl response as the rope is pulled. If necessary, replace the star brake, retainer cup, and brake spring.

Figure 4-12 shows top-of-the-line Eaton starter used on industrial engines. Service procedures are slightly more complex than for lighter-duty units because the sheave is split. This makes rope replacement more difficult, and the mainspring, which is not held captive in a retainer, can thresh about when the sheave is removed.

Disassembly

1. Remove five screws securing starter assembly to blower housing.

2. Release spring preload. Most h-d models employ a notched sheave that allows rope slack for disarming (see Fig. 4-2).

3. Remove retainer screw and any washers that may be present.

4. Lift off clutch assembly, together with brake spring and optional brake spring washer.

5. Carefully extract sheave, keeping mainspring confined

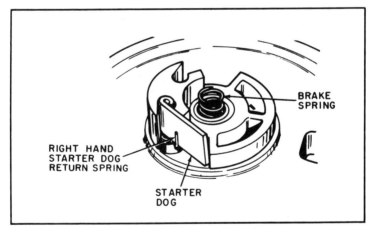

Fig. 4-11. Eaton rewind starter partially dissembled. Generous retainer screw torque compresses brake spring, generating friction against retainer that extends dog. Because the rope attaches to the inboard—and accessible—side of the sheave, the rope can be replaced by applying and holding mainspring pre-tension as shown. Original rope is fished out, new rope is passed through the eyelet and sheave hole, knotted, and pretension is slowly released. Spring winds rope over sheave.

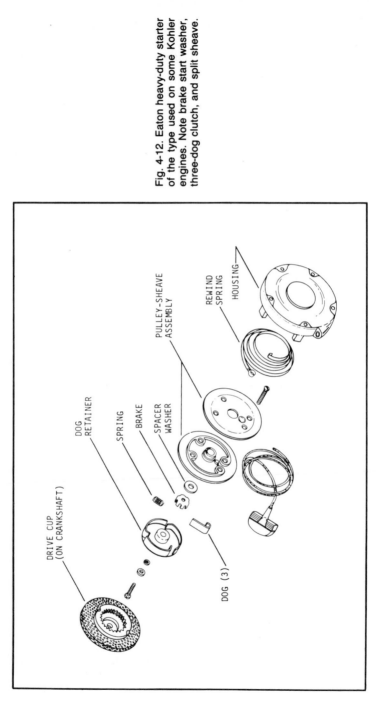

Fig. 4-12. Eaton heavy-duty starter of the type used on some Kohler engines. Note brake start washer, three-dog clutch, and split sheave.

DRIVE CUP (ON CRANKSHAFT)

DOG RETAINER

SPRING

BRAKE

SPACER WASHER

PULLEY-SHEAVE ASSEMBLY

REWIND SPRING

HOUSING

DOG (3)

within starter housing. Wear safety glasses during this and subsequent operations.

6. Remove rope, which may be knotted on the inboard side of sheave or which may be sandwiched between sheave halves as shown in Fig. 4-12. Screws that hold sheave halves together might require a hammer impact tool to loosen.

7. Remove the spring if it is to be replaced. Springs without a retainer are unwound a coil at a time from the center outward.

8. Clean and inspect with particular attention to the clutch mechanism. Older light-duty and medium-duty models employed a shouldered clutch retainer screw with a 10-32 thread. This part can be updated to a 12-28 thread (Tecumseh part No. 590409A) by retapping the sheave pivot shaft.

Assembly

1. Apply a light film of grease to the mainspring and sheave pivot shaft. Do not over-lubricate because the brake spring and clutch assembly must be dry to develop engagement friction. Snow-proof clutches, recognized by application and by half-moon pawl cam, might benefit from a few drops of oil on the pawl posts.

2. Install the rewind spring. Loose springs are supplied in a disposable retainer clip. Position the spring—observing correct engine rotation as illustrated in Fig. 4-3—over the housing anchor pin. Gently cut the tape holding the spring to the retainer, retrieving tape in segments. Use spring and retainer sets by simply dropping them in place.

3. Install rope, an operation that varies with sheave construction.

Split Sheave

A) Double knot rope, cauterize and install between sheave halves—trapping tope in cavity provided.

B) Install sheave on sheave pivot shaft, engaging inner end of mainspring. A punch or piece of wire can be used to snag the spring end as shown in Fig. 4-13. Install clutch assembly.

C) Wind sheave until mainspring coil binds (Fig. 4-14).

D) Carefully release spring tension two revolutions and align rope end with eyelet in starter housing.

E) Using Vise-Grips, clamp sheave to hold spring tension and guide rope through eyelet. Attach handle.

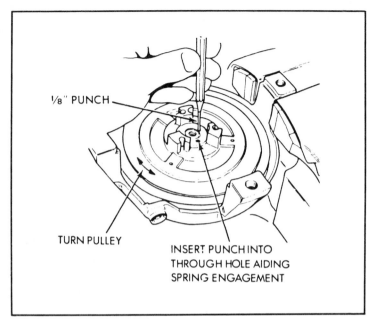

1/8" PUNCH

TURN PULLEY

INSERT PUNCH INTO
THROUGH HOLE AIDING
SPRING ENGAGEMENT

Fig. 4-13. A punch aids spring-to-sheavge engagement on large Eaton starters.

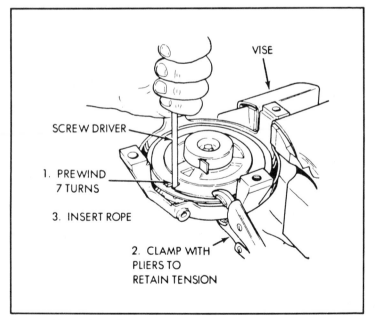

VISE

SCREW DRIVER

1. PREWIND
   7 TURNS

3. INSERT ROPE

2. CLAMP WITH
   PLIERS TO
   RETAIN TENSION

Fig. 4-14. Prewind specification varies with starter model and mainspring condition.

132

F) Verify that sufficient pretension is present to retract rope.

One-Piece Sheave

A) Wind sheave to coil bind and back off to align rope hole on inboard face of sheave with housing eyelet.

B) Clamp sheave.

C) Cauterize ends of ropes and install rope through eyelet and sheave (Fig. 4-15).

D) Knot rope under sheave and install handle.

E) Carefully release sheave, allowing rope to wind as spring relaxes.

F) Test for proper pretension.

3. Pull out centering pin (where fitted) so that it protrudes about 1/8 inch past the end of the clutch retainer screw. Some models employ a centering pin bushing.

4. Install starter assembly on engine, pulling the starter through several revolutions before hold-down screws are snubbed. Test operation.

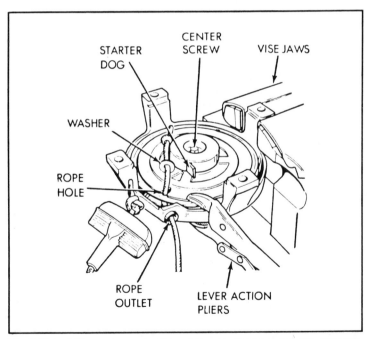

Fig. 4-15. Installing rope on one-piece sheave involves passing rope from outside starter housing, through eyelet, and into sheave connection point.

Fairbanks-Morse

Fitted to several American engines, F-M starters can be recognized by absence of serrations on the flywheel cup. The cup is a soft aluminum casting and friction shoes (clutch or brake shoes) are sharpened for purchase. Early models used a wireline in lieu of the rope. Figure 4-16 is a composite drawing of Models 425 and 475, intended for larger single-cylinder engines.

Disassembly

1. Remove starter assembly from blower housing.

2. Turn starter over on bench and, holding the large washer down with thumb pressure, remove the retainer ring that secures the sheave and clutch assembly (A of Fig. 4-17).

3. Remove the washer, brake spring and friction shoe assembly. Normally, friction shoe assembly is not broken down further.

4. Relieve mainspring preload. This can be accomplished by removing the rope handle and allowing the sheave to unwind in a controlled fashion. Tension on Model 475 can be released by removing screws holding middle and mounting flanges together (B of Fig. 4-1).

5. Cautiously lift the sheave about 1/2 inch out of housing and detach the inner spring end from the sheave hub.

6. The mainspring is left undisturbed (unless it is to be replaced). From the center outward, remove a coil at a time.

7. Clean parts in solvent and inspect.

Assembly

1. Install spring, hooking spring eyelet over anchor pin on cover. The spring lay shown in D of Fig. 4-17 is for conventional—clockwise when facing flywheel—engine rotation.

2 Rope installation and preload varies with starter model. In all cases, rope is attached to sheave and wound on it before the sheave is fitted to starter cover and mainspring. Model 475 employs a split rope guide, or ferrule, consisting of a notch in the middle flange and starter housing. Consequently, rope may be secured to and wound overflange with rope handle attached. Model 425 and most other F-M starters use a one-piece ferrule and rope must be installed without a handle. After the sheave is secured and the preload established, the rope is threaded through the ferrule for handle attachment.

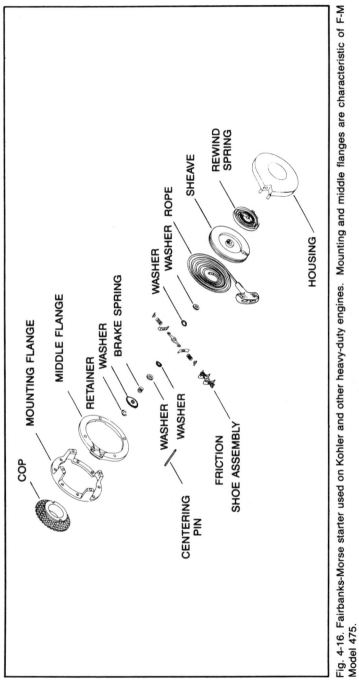

COP

MOUNTING FLANGE

MIDDLE FLANGE

RETAINER

WASHER

BRAKE SPRING

WASHER

WASHER ROPE

SHEAVE

REWIND SPRING

HOUSING

CENTERING PIN

WASHER

WASHER

FRICTION SHOE ASSEMBLY

Fig. 4-16. Fairbanks-Morse starter used on Kohler and other heavy-duty engines. Mounting and middle flanges are characteristic of F-M Model 475.

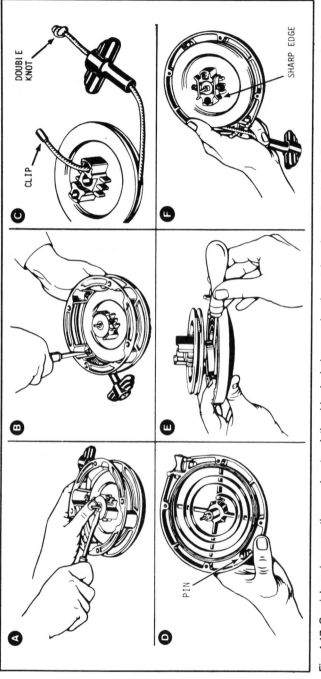

Fig. 4-17. Crucial service operations and parts relationships include removing the retainer ring and spring-loaded washer (A), releasing residual spring tension, flanged-starter shown (B), rope anchors and rope lay, standard engine rotation (C), mainspring orientation, standard rotation (D), engaging spring and sheave (E), and correct brake shoe assembly (D).

3. Lubricate the sheave pivot shaft with light grease and apply a small quantity of motor oil to the mainspring. Avoid overlubrication.

4. Install the sheave over the sheave pivot shaft with the rope fully wound. Using a screwdriver, hook the inner end of spring into sheave hub (E of Fig. 4-17).

5. Establish preload, four sheave revolutions against the direction of engine rotation for Model 425, five turns for Model 475 and variable for others.

6. Complete assembly, installing sheave hold-down hardware and friction shoe assembly. When assembled correctly, sharp edges of friction shoes are poised for leading contact with flywheel hub ID (F of Fig. 4-17).

7. Pull centering pin out about 1/8 of an inch for positive engagement with the crankshaft centerhole.

8. Install the assembled starter on the flower housing, rotating the flywheel with the starter rope as hold-down screws are torqued. This procedure helps to center the clutch in the flywheel hub.

9. Start the engine to verify starter operation.

The Fairbanks-Morse utility starter is a smaller and simpler version of the heavy-duty models just discussed. The starter housing mounts directly to the engine cooling shroud (eliminating flanges). A one-piece sheave is used with the rope, anchored by a knot, rather than compression fitting. The utility starter uses the same clutch components as its larger counterparts and, like them, can be assembled for right-hand or left-hand engine rotation. See Fig. 4-18.

VERTICAL PULL

Like other spring-powered devices, vertical-pull starters must be disarmed before disassembly. Otherwise, the starter will disarm itself with unpredictable results. Disarming involves three distinct steps: releasing mainspring pretension (usually by slipping a foot or so of rope out of the sheave flange and allowing the sheave to unwind), disengaging mainspring anchor (usually held by a threaded fastener) and, when the spring is to be replaced, uncoiling the spring from its housing. Safety glasses are mandatory.

Vertical-pull starters tend to be mechanically complex and—because of a heavy reliance upon plastic, light-gauge steel, and spring wire—are unforgiving. Parts easily bend or break. Lay components out on the bench in proper orientation and in sequence

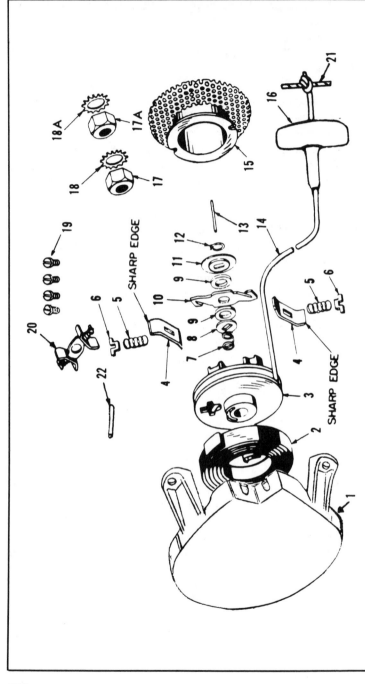

SHARP EDGE

SHARP EDGE

ILLUS. NO.	QTY.	DESCRIPTION
1	1	Cover
2	1	Rewind spring
3	1	Rotor
4	2	Friction shoe plate
5	2	Friction shoe spring
6	2	Spring retainer plate
7	1	Brake spring
8	1	Brake washer
9	2	Fiber washer
10	1	Brake lever
11	1	Brake retainer washer
12	1	Retainer ring
13	1	Centering pin
14	1	Cord
15	1	Cup and screen
16	1	T-handle
17	1	L.H. thick hex nut
17A	1	R.H. thick hex nut
18	1	Ext. tooth lockwasher (left hand)
18A	1	Ext. tooth lockwasher (right hand)
19	4	Pan hd. screw w/int.-ext. tooth lockwasher
20	1	Friction shoe assembly, includes: Items 4, 5, 6 and 10
21	1	Spiral pin
22	1	Roll pin

Fig. 4-18. Small series Fairbanks-Morse employs integral cover and mounting pedestals. Knot at the rope sheave end can be set up for right or left rotation. This example is used by Chrysler L series engines.

of disassembly. If there is any likelihood of confusion, make sketches to guide assembly. Also note that step-by-step instructions in this book must aim at thoroughness and describe all operations. It will rarely be necessary to follow every step and to completely dismantle a starter.

Briggs & Stratton

Briggs & Stratton uses one vertical-pull starter with minor variations in the link and sheave mechanisms. It is probably the most reliable of these starters and the easiest to repair.

Disassembly

1. Remove starter assembly from engine.

2. Release mainspring pre-tension by lifting the rope out of the sheave flange and, using the rope for purchase, winding the sheave counterclockwise two or three revolutions (Fig. 4-19).

3. Carefully pry the plastic cover off with a screwdriver. Do not pull on the rope with the cover off and spring anchor attached. Under these conditions it is possible for the outer end of the spring to slip out of the housing.

4. Remove the spring anchor bolt and spring anchor (Fig. 4-20). If the mainspring is to be replaced, carefully extract it from the housing, working from center coil outward. Note the spring lay for future reference.

5. Separate the sheave and pin (Fig. 4-21). Observe the link orientation.

6. Rope can be detached from sheave with the aid of long-nosed pliers. Figure 4-22 illustrates this operation and link retainer variations.

7. Rope can be disengaged from handle by prying the handle center section free and cutting the knot (Fig. 4-23).

8. Clean parts (except rope) in petroleum-based solvent to remove all traces of lubricant.

9. Verify gear response to link movement as shown in Fig. 4-24. The gear should move easily between its travel limits. Replace link as necessary.

Assembly

1. Install the outer end of mainspring in housing retainer slot and wind counterclockwise (Fig. 4-25).

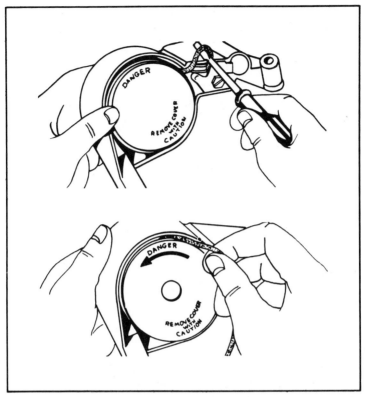

Fig. 4-19. Briggs & Stratton vertical-pull starters are disarmed by slipping rope out of sheave groove and using the rope to turn the sheave two or three revolutions counterclockwise until the mainspring relaxes.

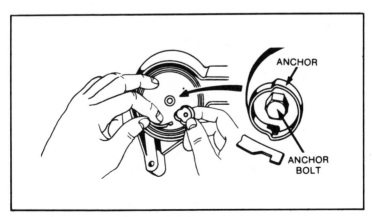

Fig. 4-20. Mainspring anchor bolt must be torqued 75-90 inch-pound and can be further secured with thread adhesive.

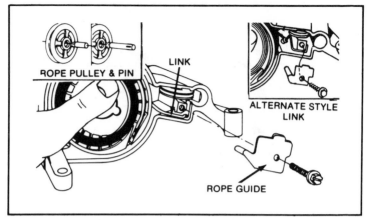

Fig. 4-21. Observe friction link orientation for assembly.

2. Guide the rope into the sheave with the help of a small screwdriver or stiff wire (Fig. 4-26).

3. Knot the rope and sear the end in the open flame to prevent unraveling. Pull the rope hard; the seating knot must not interfere with link operation.

4. Install rope in handle.

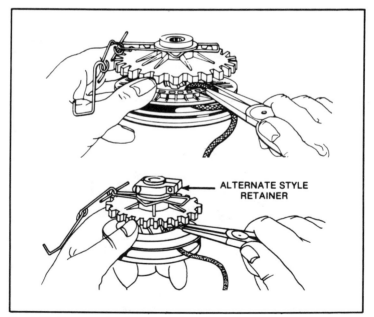

Fig. 4-22. Rope can be disengaged from sheave with long-nosed pliers.

142

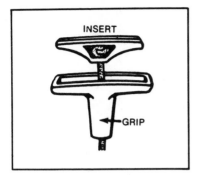

Fig. 4-23. Briggs & Stratton handle insert must be pried out of grip for rope installation.

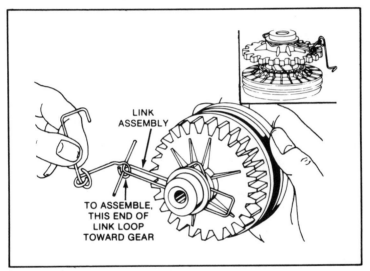

Fig. 4-24. Pinion gear should move through its full range of travel in response to link movement. Note orientation of link for assembly (inset).

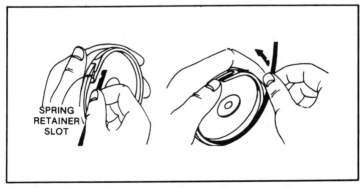

Fig. 4-25. Mainspring winds counterclockwise from outer coil.

143

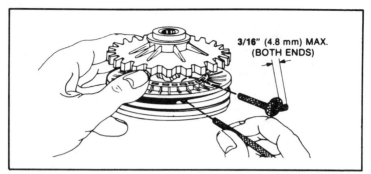

Fig. 4-26. A short length of piano wire aids rope insertion into sheave.

5. Mount the sheave, sheave pin and link assembly in the housing. Index the end of link in groove or hole provided (Fig. 4-27).

6. Install the rope guide and hold-down screw.

7. Rotate the sheave counterclockwise, winding the rope over sheave (Fig. 4-28).

8. Engage the inner end of the mainspring on the spring anchor. Mount anchor and torque hold-down capscrew 75-90 inch-pounds.

9. Snap plastic cover into place over spring cavity.

10. Disengage 12 inches or so of rope from the sheave and, using rope for purchase, turn the sheave two or three revolutions clockwise to generate pre-tension. See Fig. 4-29.

11. Mount starter on engine and test.

Tecumseh

Tecumseh has used several vertical-pull starters, ranging from

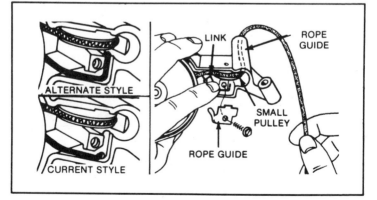

Fig. 4-27. Friction link hold-down detail.

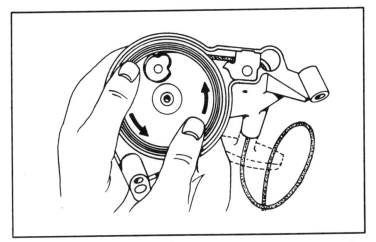

Fig. 4-28. The rope winds counterclockwise on the sheave, and then the spring anchor and anchor bolt are installed.

quickie adaptations of side-pull designs in the 1960s to the current vertical-engagement type, which stands as a kind of textbook example of modern engineering and manufacturing techniques.

The *gear-driven starter* shown in Fig. 4-30 is an interesting transition from side to vertical-pull. No special service instructions seem appropriate, except to provide plenty of grease in the gear housing and some light lubrication on the mainspring. Assemble the brake spring without lubricant.

The current *horizontal-engagement starter* (Fig. 4-31) is reminiscent of the Briggs & Stratton design, with rope clip, cup-

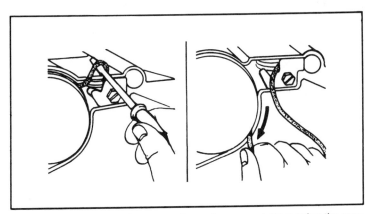

Fig. 4-29. Pretension requires two or three sheave revolutions using the rope for leverage.

145

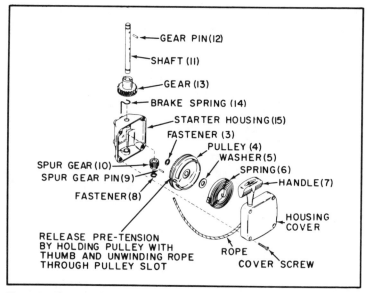

Fig. 4-30. Early Tecumseh vertical-pull starter, driving through a gear train. While heavy and, no doubt, expensive to manufacture, this starter was quite reliable.

type spring anchor ("hub" in the drawing) and threaded sheave extension upon which the pinion rides.

Disassembly

1. Remove the unit from the engine.

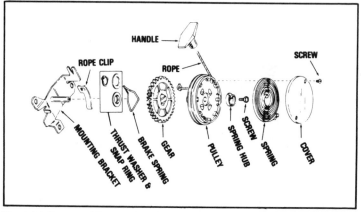

Fig. 4-31. Tecumseh's most widely used vertical-pull starter employs a spiral gear to translate the pinion horizontally into contact with the flywheel.

2. Detach the handle and allow the rope to retract past the rope clip. This operation relieves mainspring preload tension.

3. Remove two screws, while holding the cover in place, and carefully pry the cover free.

4. Remove the hold-down screw and hub.

5. Extract mainspring from housing, working a coil at a time from the center out. If it will be reused, the spring can be left undisturbed.

6. Lift off the gear and pulley assembly. Disengage the gear and, if necessary, remove the rope from the pulley.

7. Clean parts.

8. Inspect the friction spring (the Achilles' heel of vertical-pull starters). The spring must be in solid contact with the groove in gear.

Assembly

1. Secure the rope to the handle, using No. 4 1/2 or 5 nylon rope, 61 inches long for standard starter configurations. Sear rope ends and form by wiping with a cloth while the rope is still hot.

2. Assemble gear on pulley, using no lubricant.

3. Lightly grease center shaft and install gear and pulley. Brake spring loop is secured by bracket tab. Rope clip indexes with hole in bracket (Fig. 4-32).

4. Install hub and torque center screw to 44-55 inch-pound.

5. Install spring. New springs are packed in a retainer clip to make installation easier.

6. Install cover and cover screws.

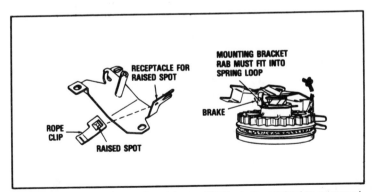

Fig. 4-32. Generous gear lash, minimum 1/16 inch, is required to assure pinion disengagement when engine starts.

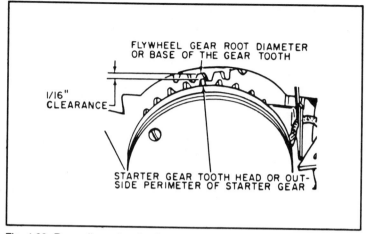

FLYWHEEL GEAR ROOT DIAMETER
OR BASE OF THE GEAR TOOTH

1/16"
CLEARANCE

STARTER GEAR TOOTH HEAD OR OUT-
SIDE PERIMETER OF STARTER GEAR

Fig. 4-33. Rope clip and spring loop index to bracket.

7. Wind rope on pulley by slipping it past rope clip. When fully wound, turn pulley two additional revolutions for preload.

8. Mount starter on engine, adjusting bracket for minimum 1/16-inch tooth clearance (Fig. 4-33). Less clearance could prevent disengagement, grenading the starter.

The *vertical-pull, vertical-engagement* starter is a serious piece of work that demands special service procedures. It is relatively easy to disassemble while still armed. Results of this error can be painful. Another point to note is that rope-to-sheave assembly as done in the field varies from the original factory assembly.

Figure 4-34 is a composite drawing of several vertical-pull starters. Many do not contain the astericked parts, and early models do not have the V-shaped groove on the upper edge of the bracket that simplifies rope replacement.

When this groove is present, the rope (No. 4 1/2, 65-inch standard length, longer with remote rope handle) can be renewed by turning the sheave until the staple, which holds the rope to the sheave, is visible at the groove (Fig. 4-35). Pry out the staple and wind the sheave tight. Release the sheave just some 180° to index the hole in the sheave with the V-groove. Insert one end of the replacement rope through the hole, out through the bracket. Cauterize and knot the short end, and pull the rope through, burying the knot in the sheave cavity. Install rope handle, replacing the original staple with a knot, and release the sheave. The rope should wind itself into place.

Disassembly

1. Remove starter from engine.

2. Pull out the rope far enough to secure rope in the V-wedge on the bracket end. This part, distinguished from the V-groove mentioned above, is called out in Fig. 4-34.

3. The rope handle can be removed by prying out the staple with a small screwdriver.

4. Press out the head pin that supports the sheave and spring the capsule in the bracket. This can be done in a vise with a large deepwell wrench socket as backup.

5. Turn the spring capsule to align with brake spring legs. Insert a nail or short (3/4-inch maximum) pin through the hole in strut and into gear teeth (Fig. 4-36).

6. Lift the sheave assembly and spring the capsule out of the bracket.

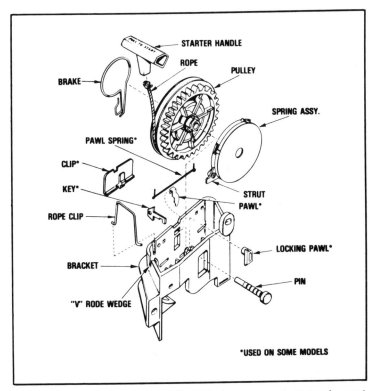

Fig. 4-34. Tecumseh's vertical-pull, vertical-engagement starter is most sophisticated unit used on small engines. Spring and cover are integral and are not separated for service.

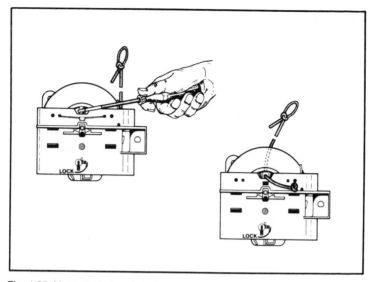

Fig. 4-35. V-groove in bracket gives access to rope anchor on some models.

Warning: Do not separate the sheave assembly and spring capsule until the mainspring is completely disarmed.

7. Hold the spring capsule firmly against the outer edge of

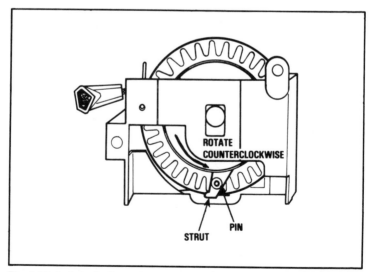

Fig. 4-36. A pin locks the spring capsule and gear to prevent sudden release of mainspring tension.

the sheave with thumb pressure and extract the locking pin inserted in Step 5.

8. Relax pressure on the spring capsule, allowing the capsule to rotate, and dissipating residual mainspring tension.

9. Separate the capsule from sheave and, if rope replacement is in order, then remove hold-down staple from sheave.

10. Clean and inspect parts.

Note: No lubricant is used on any part of this starter.

Assembly

1. Cauterize and form ends of replacement rope (see specs above) by wiping down with a rag while still hot.

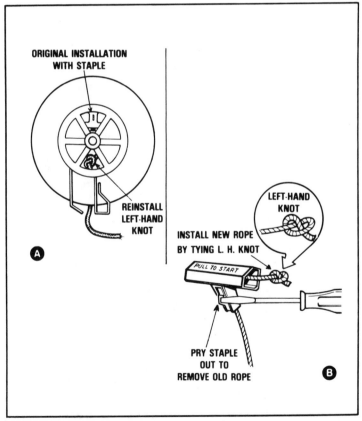

ORIGINAL INSTALLATION
WITH STAPLE

REINSTALL
LEFT-HAND
KNOT

A

LEFT-HAND
KNOT

INSTALL NEW ROPE
BY TYING L. H. KNOT

PULL TO START

PRY STAPLE
OUT TO
REMOVE OLD ROPE

B

Fig. 4-37. Replacement rope anchors with knot, rather than staple, and mounts 180° from original position on sheave.

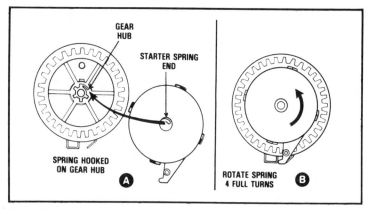

Fig. 4-38. Spring capsule engages gear hub (A), is rotated four revolutions and pinned (B).

2. Insert one end of rope into sheave, 180° away from original (staple) mount (A of Fig. 4-37).

3. Tie knot and pull rope into knot cavity.

4. Install handle (B of Fig. 4-37).

5. Wind rope clockwise—as viewed from gear—on sheave.

6. Install brake spring, spreading spring ends no more than necessary.

7. Position the spring capsule on the sheave, making certain the mainspring end engages the gear hub (A of Fig. 4-38).

8. Wind four revolutions, align the brake spring ends with the strut (B of Fig. 4-38), and lock with the pin used during disassembly.

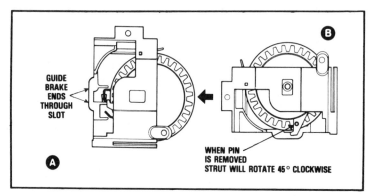

Fig. 4-39. Sheave and spring capsule assembly installs in bracket with brake spring ends in slots (A). Releasing pin arms starter (B), which can now be mounted on the engine.

9. Install pawls, springs, and other hardware that might be present.

10. Insert sheave and spring assembly into bracket, with brake spring legs in bracket slots (Fig. 4-39).

11. Feed rope under guide and snub in V-notch.

12. Remove locking pin, allowing strut to rotate clockwise until retained by bracket.

13. Press or drive center pin home.

14. Mount starter on engine and test.

Chapter 5

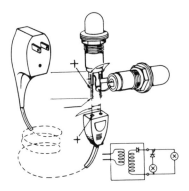

Electrical System

At its most developed, the electrical system consists of a charging circuit, or voltage source, and a starting circuit, or primary load (Figs. 5-1 and 5-2). Secondary loads, such as lights and instrumentation, might also be present. An alternator or generator are the primary elements of the charging circuit, (which also can contain a regulator, rectifier and fuse). The starting circuit includes a starter motor, solenoid or relay, and—in nearly all examples—a battery.

Not all small-engine electrical systems include both circuits. Some dispense with the starting circuit and others employ a starting circuit without provision for onboard power generation.

STARTING CIRCUITS

Starting circuits fall into two groups: dc (direct current) systems that are powered by 12 or 6V batteries, and ac (alternating current) systems that are fed directly through a 120 Vac line. Most dc systems employ lead-acid batteries that, with rare exceptions, are replenished by an engine-driven charging circuit. Nicad batteries must be recharged before use and so depend upon household current to provide the necessary energy.

Batteries

Outright battery failure—zero cranking voltage after

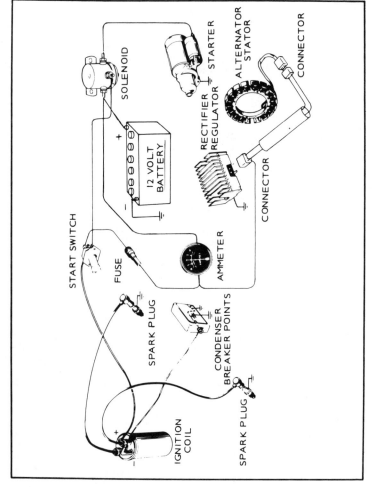

Fig. 5-1. Electrical system used on Onan engines ties into coil and battery ignition. Starter operates by way of solenoid.

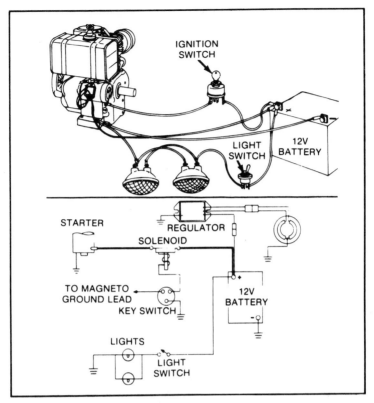

Fig. 5-2. Briggs & Stratton 10A system combines magneto ignition with accessory tap for lights.

charging—is no problem to troubleshoot. But partial failure, characterized by some loss of battery capacity, can be more difficult to pinpoint. The battery is the connection point between charging and starting circuits, and either circuit could be at fault. However, the battery is usually the culprit and the first order of business is to charge the battery and test its output.

Lead-Acid. The lead-acid storage battery is the most popular type. It can be recharged several hundred times and maintains its charge fairly well, declining by 1 percent or 2 percent a day. On the other hand, lead-acid batteries lose capacity when cold, dropping to less than 70 percent of normal at 32 °F. and, even at best, are able to convert only a tiny fraction of charging current to stored energy. These batteries contain fairly concentrated solutions of sulfuric acid and give off hydrogen gas, especially during charging. The combination of a corrosive liquid and an explosive gas is hazardous. **Warning:** Wear safety glasses when servicing a lead-acid battery. To prevent sparks, disconnect the battery charger before charger-to-battery connections are made or broken. Test as follows:

1. Open battery-cable connections and scrape connectors and terminals to bright metal. Check cable-to-ground and cable-to-starter connections. Remove any rust, corrosion, or paint deposits present. Secure all cable connections.

2. Verify that electrolyte level covers plates in each cell. Secure the caps.

3. Clean battery top with a mixture of baking soda, detergent, and water. Rinse with fresh water and wipe dry.

4. Charge the battery, using a taper charger (i.e., a regulated charger that reduces output as battery approaches full charge). Note that charging releases hydrogen gas.

5. Fill battery cells with distilled water (as necessary).

6. Test each cell with a temperature-compensated hydrometer (Fig. 5-3). Replace the battery if hydrometer readings average less than 1.225 or if individual cell readings vary by more than 0.050.

7. Connect voltmeter across battery terminals (Fig. 5-4).

Fig. 5-3. Use a hydrometer to determine specific gravity of each battery cell: 100% = 1.260 to 1.280; 75% charge = 1.230 to 1.250; 5-% charge = 1.200 to 1.220; 25% charge = 1.170 to 1.190; complete discharge = 1.110 to 1.130. Specific gravity readings are referenced to 80°F electrolyte temperature. Add 0.004 for each 10° F over 80° and subtract 0.004 for each 10° below 80°.

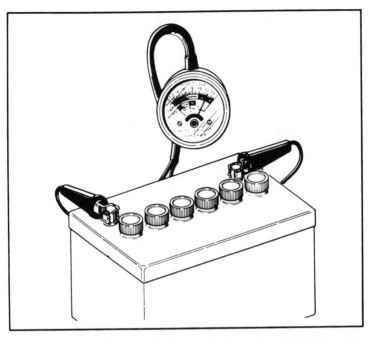

Fig. 5-4. A fully charged 12V battery should deliver at least 9.5V during a 10- to 15-second discharge through the starter. A 6V battery should maintain a minimum of 4.5V during the same test. Lower than specified cranking voltages usually can be traced to the battery, although the starter motor, its wiring and even parasitic engine loads can increase the current draw and cost voltage.

8. Disable the ignition, energize the starter motor, and observe the voltmeter.

Caution: Small-engine starter motors are intended for brief, intermittent service and must not be operated for more than 10 seconds continuously. Allow 60-second cool-down period between each 10-second cranking episode.

9. Cranking voltage should remain above 9.5V (12V systems) or 4.5V (6V). Lower readings mean a defective battery or starter circuit.

Nicad. Nickel-cadmium battery packs are supplied by Briggs & Stratton, Tecumseh, Toro, and other manufacturers as part of a starting system that consists of 12V motor, switch, and 120Vac battery charger (Fig. 5-5). These batteries are based on an entirely different chemistry than conventional storage batteries and require special handling and maintenance procedures. Other elements of the system do not interchange with more conventional hardware. An ordinary battery charger cannot be used to recharge nicad batteries, nor can a standard 12V starter or automotive ignition switch be easily substituted.

Warning: Nicad batteries have potential for harm. Basic precautions include extreme care in handling and disposal. Cadmium, visible as a white powder on leaking cells, is a persistent poison. Do not incinerate and avoid welding on or around the battery pack.

While the charger can be plugged in continuously, battery (and charger) life will be extended if charging is limited to a 12-hour or 16-hour period just prior to use. This should fully replenish the battery pack with energy for 30 to 40 engine starts. Deep discharge should also be avoided. When the unit is in storage, batteries should be charged for 12 to 16 hours once every two months. Do not charge when ambient temperature drops below 40° F (4° C).

Weak starter response when connected to a freshly charged battery pack can mean any of the following problems:

- Faulty battery charger.
- Defective starter motor or switch circuit.

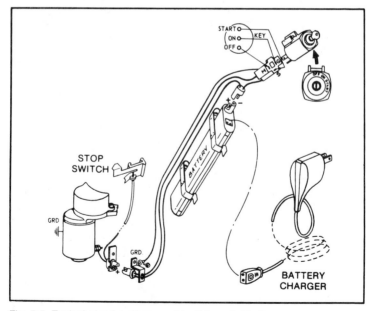

Fig. 5-5. Typical nicad system used by Briggs & Stratton.

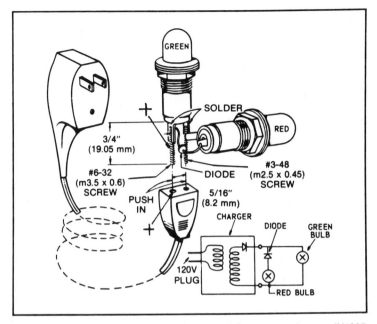

Fig. 5-6. This charger tester, designed by Briggs & Stratton, requires one IN4005 diode, two Dialco lamp sockets (red No. 0931-102 and green No. 0932-102) two No. 53 bulbs, one 6-32, 3/4-inch screw and one 3-48, 3/4-inch screw. These, or equivalent, parts are available at electronic supply stores. A battery charger in good condition will light the green lamp only. A charger with an open diode will light the red bulb and one with a shorted diode will light both bulbs.

- Engine bind.
- Defective battery pack.

In general, if the charger becomes warm when plugged in, it is operating properly. Output is variable, depending upon battery charge, but after two or three hours the battery should draw something on the order of 80 mA. Tecumseh supplies a test meter—part No. 670235—for the 32659 Nicad charger. I do not know whether this meter can be used with competitive chargers. Briggs & Stratton suggests that the technician construct his own tester, to indicate whether the charger diode is open or shorted (Fig. 5-6).

Parasitic loads are detected by turning the flywheel by hand. The usual cause is a dragging brake or clutch. More ominous possibilities include hydrostatic lock and wiped engine bearings.

The most common fault is refusal of the Nicad battery to take a full charge. After 16 hours on the charger, battery potential should register no less than 15.5V and should not be greater than 18V.

Assuming that static (motor switched off) voltage is within these limits, the next step is to measure battery capacity. This amounts to a *controlled* discharge, achieved by shunting output leads across a load. One technique, suggested by Tecumseh, is to connect a 1.4 ohm (± 10 percent), 150W resistor across the output. With the starter disconnected, discharge the battery through the resistor for 2 minutes. Check output voltage with a 10,000-20,000 ohm/volt voltmeter. Battery voltage at the end of the test should be at least 9V.

Another approach is to use two No. 4001 sealed-beam headlamps with the terminals soldered together as shown in Fig. 5-7. When connected to a fresh battery pack, bulbs should burn brightly for 6 minutes (Tecumseh) or 5 minutes (Briggs & Stratton). Briggs suggests that a voltmeter, connected across the two lamp terminals, should show 13.5V minimum after 1 minute—with 13V or less indicating a shorted cell.

These tests, whether using a resistor or headlamps, are one-

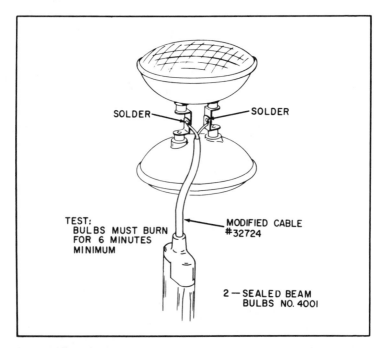

Fig. 5-7. Perhaps the easiest way to check nicad output is to discharge the battery pack through two headlamps. The hookup shown uses a modified battery to starter motor cable.

shot affairs and cannot be repeated until the battery is charged for another 14 to 16 hours.

Starter Motors

A mechanic can expect to encounter a variety of starter motors, sometimes interchangeable between engine makes and models, but more often specific to the engine at hand. Major manufacturers include American Bosch, European Bosch, Bendix, Briggs & Stratton, Mitsuibishi, Motor Products, and Tecumseh. Motors fall into the following three basic groups:

dc—Lead-Acid Battery. Rated at 6V to 12V, these motors are generally two-pole (brush) designs with electromagnetic fields. Figure 5-8 illustrates one type, differing from most only by the use of tubular insulators over the through-bolts. Figure 5-9 shows a Bendix starter, used on a variety of engines, including some of Japanese manufacture. Figure 5-10 shows yet another variation. While some early motors employed solenoid, or linear motor drive, most modern designs rely upon an inertial clutch, or "Bendix," to engage the pinion with the flywheel.

dc—Nicad Battery. Representing a later generation than lead-acid types, these 12V units typically employ permanent-magnet fields, end-cap brush mounting and individualized housings, designed to fit a narrow range of engines. Figure 5-11 show a Nicad starter in various stages of disassembly.

ac —External Power Supply. The distinguishing feature of 120Vac motors is a rectifier mounted on the housing that converts line voltage to pulsating dc, which registers on a meter as about 15V less than the ac input. A four-pole, permanent-magnet field unit for large displacement engines is shown in Fig. 5-12. This starter is also available in a 12V version. Note that the Bendix mechanism is secured with a compression pin (Fig. 5-13).

Preliminary Tests. Turn the flywheel through several revolutions by hand to verify that the engine is free. Using heavy jumper cables, connect a "hot" battery to the motor terminal and an engine ground, observing correct polarity. Control circuitry for 120Vac motors can be shunted in the same manner, using an extension cord. However, this procedure involves extreme and potentially lethal electroshock hazard, and I do *not* recommend it. If the motor is shorted, the engine and whatever the engine is attached to will be hot.

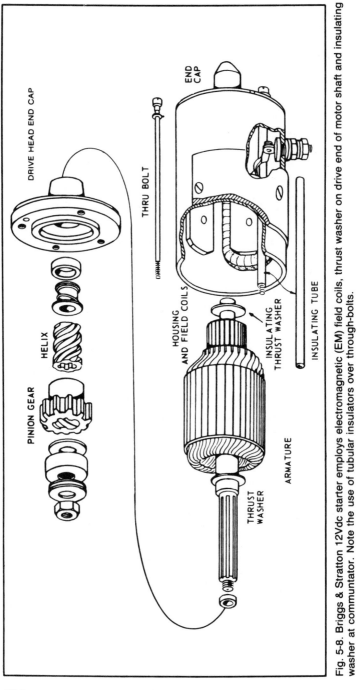

PINION GEAR

HELIX

DRIVE HEAD END CAP

THRU BOLT

END CAP

HOUSING AND FIELD COILS

INSULATING THRUST WASHER

INSULATING TUBE

ARMATURE

THRUST WASHER

Fig. 5-8. Briggs & Stratton 12Vdc starter employs electromagnetic (EM) field coils, thrust washer on drive end of motor shaft and insulating washer at communtator. Note the use of tubular insulators over through-bolts.

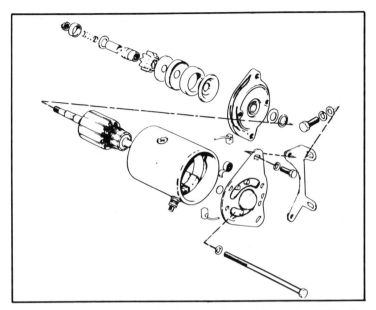

Fig. 5-9. Bendix 12Vdc motor with thrust washer at drive end and EM coils. This motor is used on a number of American and foreign-made engines.

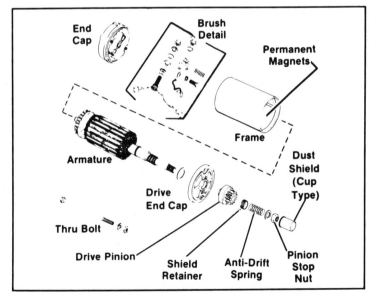

Fig. 5-10. American Bosch 12Vdc starter used on Kohler and other large engines employs permanent magnet (PM) fields and radial communtator with brushes parallel to motor shaft.

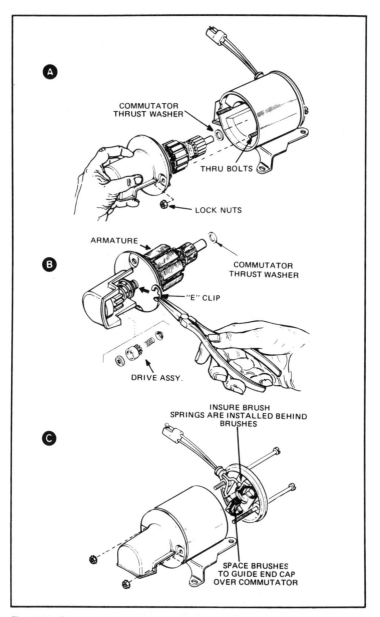

Fig. 5-11. Tecumseh nicad end caps are assembled on motor housing with two through-bolts (A). Bendix drive assembly is secured on shaft with E-clip (B). Brushes, which are replaceable only as part of the end cap assembly, must be carefully shoehorned over commutator (C). Thrust washer mounts on commutator end of motor shaft. This particular motor should draw 20A while cranking the engine 415 rpm (with lube oil at 70° F).

166

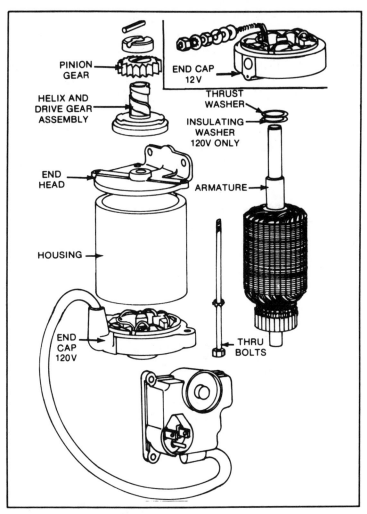

Fig. 5-12. Briggs & Stratton 120Vac starter used on 17-through-32-cubic inch engines. (A 12V version is available.) The insulating washer is a safety feature that must be installed as shown.

A healthy starter should spin the flywheel 350 rpm or so with the spark plug installed. Possible malfunctions include:

● Motor runs free without engaging flywheel. Check Bendix drive for dirt accumulations, broken parts.

● Motor runs, Bendix "machineguns" in and out. Check Bendix pinion and flywheel ring gears for interference.

● Motor runs but slowly. Check engine ignition timing (if variable), motor bearing side play, motor shaft straightness, pinion/flywheel clearance (usually specified as 1/16 inch between tip of pinion tooth and root of flywheel tooth) commentator, armature, and fields.

● Motor does not run; no spark when jumper connection is broken. Open circuit, and check brushes and brush connections.

● Motor does not run, and there is a spark when the jumper connection is broken—short circuit that, at best, may be confined to brush feed circuit. If armature or fields are involved, it is usually advisable to replace the motor.

Service. Approach starter work with discretion, disassembling the unit no further than necessary to make the repair. In addition to the usual hand tools, you should have access to a voltohmmeter and a growler to check the armature. Bushings made from sintered

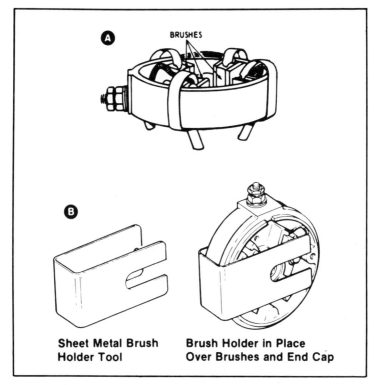

Fig. 5-13. Brush hold-down straps for four-pole unit (A) and for two-pole radial armature (B).

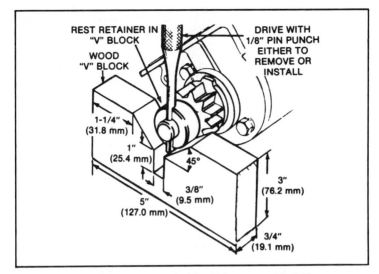

Fig. 5-14. Shaft must be supported while driving out and installing compression pin. Also note that beveled side of pinion gear—shown clearly in previous figure—must be assembled toward flywheel gear.

bronze—recognized by their cobbley appearance—require 20W motor oil. Conventional "hard" bushings like a light high-temperature grease such as Lubriplate. Interior parts should be cleaned with MEK or one of the aerosol products sold for electrical work. Use these chemicals with care and observe the warnings on the label. Crucial components include:

End Caps. Mark end caps and starter housing with a pin punch as an assembly aid.

Brushes. Most starter problems originate with the brushes that wear short and bind against the sides of their holders or that are propelled by insufficient spring tension. Older starters employed screw-type brush terminals for ease of replacement. Newer designs employ soldered terminals, occasionally silver soldered. A few starters, including the one shown in Fig. 5-11, employ integral brushes that must be replaced with the end cap as an assembly. Brush assembly tools are sometimes necessary (Fig. 5-13).

Clutch. Most often secured to the motor shaft with a conventional right-hand nut, a few have been built with left-hand threads. When a roll pin is used, secure the motor shaft in a Vee-block as shown in Fig. 5-14.

Bushings. Do not disturb bushings unless they are visibly worn. The Bendix-end bushing is usually in an open boss and can be driven out from behind; the commutator bushing is another matter because standard practice is to press the bushing into the closed end cap. It may be collapsed with a small chisel or, more elegantly, can be forced out with hydraulic pressure. Fill the bushing ID with heavy grease. Obtain a rod that exactly matches motor shaft diameter and hammer the rod into the bushing recess. Impacted grease will lift the bushing out of its bore.

Thrust Bearing. The motor shaft thrust passes through the commutator shoulder and into the end cap. One or more spacer washers one or the other end of shaft absorb wear and compensate for manufacturing tolerances to hold endplay in the 0.006-0.008-inch range. Bearing houses stock the necessary spacer washers.

Warning: nylon washers used on some motors are crucial and on 120Vac machines they are a safety-related item. Washers must be in good repair and installed as originally found. See Figs. 5-8 and 5-12.

Commutator. With brushes out of the circuit, use a continuity lamp or ohmmeter to verify that:

● Paired commutator bars adjacent to the brush holders have continuity. Rotate the armature to test each pair individually.

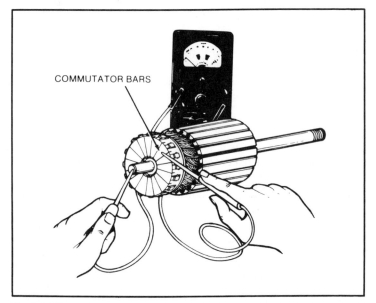

Fig. 5-15. Testing American Bosch radial commutator for shorts to motor shaft. (Courtesy Onan.)

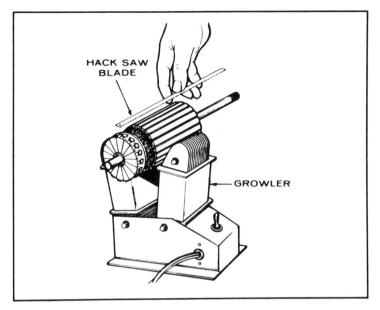

Fig. 5-16. A shorted armature windings cause thin metal strip to vibrate when assembly is magnetized on a growler. (Courtesy Onan.)

● No bar pair has continuity with other commutator bars or with the motor shaft (Fig. 5-15).

Reddish-brown discoloration on the brush track is normal and means that the brushes are seating properly. Polish out small imperfections with 000 sandpaper. Do not use emory cloth. More serious faults, including out of round, deep scores and burns can sometimes be corrected by dressing the commutator on a lathe. Afterwards, insulation should be cut out from between the segments for brush clearance with a jeweler's file. Before proceeding with these operations, make certain the commutator bars are thick enough to tolerate dressing. Many modern designs—including radial types with brushes parallel to the motor shaft—cannot be remachined.

Armature. The continuity test, described in the previous section, can be used to detect gross faults. Other indications of trouble are "thrown" solder, sometimes adhering to the motor housing ID, cooked insulation and severe pitting between adjacent commutator segments. A "growler" will respond to internal, winding-to-winding shorts that escape other tests (Fig. 5-16).

Although a moisture-shorted armature can sometimes be

revived by heating it for several hours in an oven at 250° or so, most armature faults can only be corrected by rewinding. Unfortunately, rewinding small armatures is expensive (even when you can find someone willing to do it). Purchasing a new armature is usually the best solution.

Fields. American Bosch PM fields glued in place should be inspected for signs of separation and cracks that can develop from mishandling (e.g., overtightening motor housing in a vise) or from armature contact. Electromagnetic fields generally are subject to all the ills of electrical components (Fig. 5-17), and may suffer contact with the armature if bearings are loose or motor shaft is bent. EM fields are secured by flat-head screws that usually must be loosened with an impact driver. Before dismantling, it is a good idea to check of parts availability. Replacement coils are never easy to obtain and are not always listed as separate items for recent motors. In other words, you might have to purchase a new motor housing.

Rectifier. Mounted on or near 120Vac motor housings, the rectifier assembly converts alternating line current to dc and usually incorporates the starter motor switch. Figure 5-18 shows a typical unit in schematic form. Diodes are mounted in pairs on heat sinks.

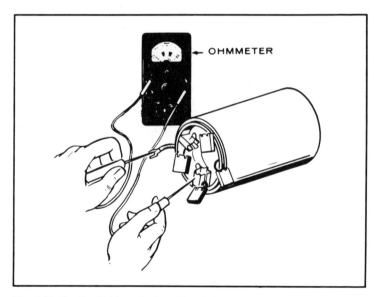

Fig. 5-17. Tracing field continuity with an ohmmeter establishes that circuits have continuity. Shorts, either to motor housing or between field windings, require special equipment to detect.

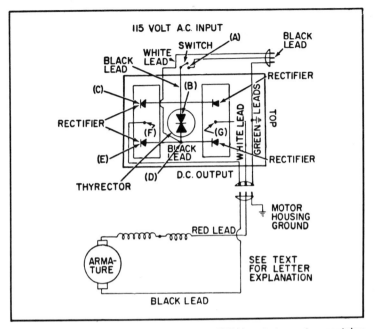

Fig. 5-18. Control box, mounted on or near 120LVac starter motor, contains motor switch, full-wave solid-state rectifier assembly and, on Tecumseh designs, a thyrector.

The thyrector is a Tecumseh feature that, while normally an insulator, conducts to ground when design voltage is exceeded in order to protect rectifiers from ac voltage surges. The unit illustrated can be opened for repair and replacement of individual components. Test rectifiers for unidirectional conductivity: connect one lead from an ohmmeter to the rectifier lead and the other lead to the housing. Note the meter reading and reverse connections. The rectifier should conduct in one direction and block current in the other. Check the thyrector to a 7.5W lamp and a 115Vac power line in series. If the lamp glows, the thyrector is bad.

Most connections are soldered and excessive heat will damage diodes and thyrector. Use alligator clips for heat sinks and monitor temperature by holding your finger to the component. If you can stand it, so can the part.

Briggs & Stratton units are partially encapsulated (although rectifiers can be replaced as a separate assembly). Figure 5-19 illustrates the basic test hookup. When the unit is working properly, it delivers no less than 14V below line voltage across a 10,000 ohm, 1W resistor. Diodes are checked as indicated in Fig. 5-20.

173

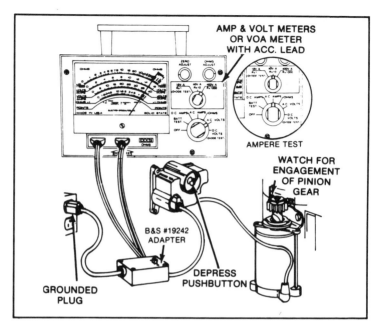

Fig. 5-19. Although this test looks complicated, it merely verifies that dc voltage to motor is no more than 14 or 15V below ac input voltage.

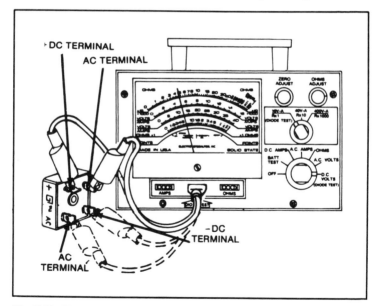

Fig. 5-20. Diodes can be checked with an ohmmeter (as illustrated here) on a Briggs & Stratton control box.

174

Solenoid and Relay. Although the Bendix-type inertial drive is by far the most popular, a few designers continue to use solenoid engagement (Fig. 5-21). A solenoid is a remotely operated linear motor that, in this application, closes contacts to provide power to the starter motor and almost simultaneously moves the pinion gear into engagement with the flywheel ring gear. Solenoid failures are usually traceable to dirt in the lever mechanism, bad control switch contacts, or bad solenoid contacts. Test the contacts by jumping battery power directly to the starter motor.

A relay is a solenoid that does not perform mechanical work (such as moving a pinion gear into mesh). In other words, a relay is merely the switching part of a solenoid. Because of the power loss and expense of long battery cables, relays are used to energize the starter motor in applications that combine Bendix drive with a remotely mounted starter switch. A relay has at least three connections: a heavy cable from the battery, a second heavy cable to the starter, and a small diameter wire to the remote switch. A second small diameter wire may be connected to ground. In any event, the test procedure is the same. Using a short length of battery cable, jump the two cable connections at the relay. The starter should energize and engage the flywheel.

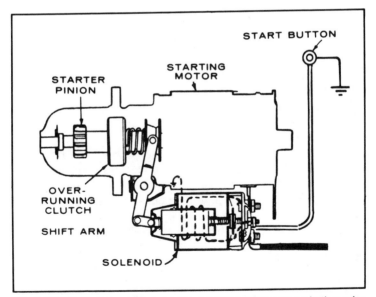

Fig. 5-21. A few starter motors employ a solenoid that acts as both a relay to energize the starter and as a linear motor to move the pinion gear into mesh with the flywheel gear.

175

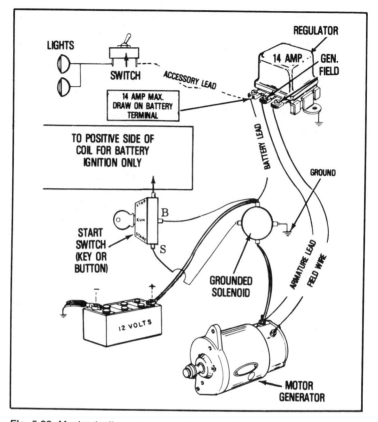

Fig. 5-22. Mechanically, motor-generators are similar to starter motors. The tricky part is the wiring.

Motor-Generator

Delco-Remy builds a belt-driven combination starter motor and generator that can be ordered as an accessory for horizontal crankshaft engines. Figure 5-22 illustrates a typical hookup that employs key starting and has provision for a optional battery ignition. With the exception of ignition, all accessory loads must be taken from the battery terminal (marked "B" or "Batt") at the voltage regulator. This particular system carries a 14A rating. With battery ignition, which draws about 3A, subtracted, 11A are available for accessories. Seven-amp systems are also encountered.

Assuming that drive belt and wiring are functional, starting problems might originate with wiring, key switch, relay (all of which may be checked with a jumper), or the motor-generator itself. Troubleshoot the latter as you would any other starter motor.

Generator problems are a little more difficult to define because both motor-generator and voltage regulator are prime suspects.

The initial test is to place an ammeter in series with the wire leading to the B regulator terminal, one meter lead to the terminal and the other to the wire, which carries charging current to the positive battery pole. Run the engine about 2000 rpm and observe the meter. Some charge should be indicated, depending upon the state of charge of the battery and the accessory load. With no charge or excessively high charge—10A or more with a healthy battery— call for a second test to discriminate between the generator and regulator.

Motor-generators have insulated fields. The unit should put out full rated amperage when the fields are grounded. Disconnect the field ("F") regulator terminal and ground the lead from the generator to the engine. Run the engine at no more than 2000 rpm. Higher speeds and resultant outputs could damage an otherwise healthy armature. If output is zero or minuscule, the motor-generator is defective. If output is 10A or more, the motor-generator can be assumed good and, almost by definition, the regulator is shorted.

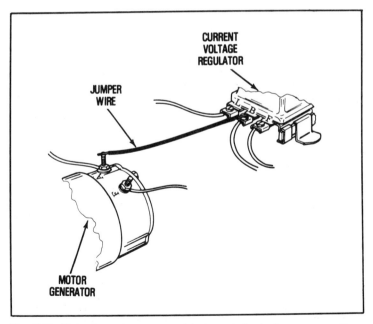

CURRENT
VOLTAGE
REGULATOR

JUMPER
WIRE

MOTOR
GENERATOR

Fig. 5-23. Motor-generator may need to be polarized when a new voltage regulator is installed or after long storage with the battery out of the circuit.

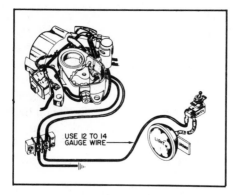

Fig. 5-24. Basic charging system consists of low voltage stator windings and load, which in this case is a headlamp.

USE 12 TO 14 GAUGE WIRE

Although the most common regulator ailment—sticking contacts—can sometimes be corrected with a point file, it is usually better to replace a faulty unit. A replacement unit must be polarized by momentarily connecting the B regulator terminal with the motor-generator armature (as shown in Fig. 5-23).

CHARGING SYSTEMS

Occasionally a mechanic will still encounter a direct current charging system, similar to the generator half of the motor-generator system shown in Fig. 5-22. However, direct-current generators have been almost universally replaced with flywheel alternators that, as the name indicated, produce alternating current. Output of the more powerful units is controlled by a solid-state regulator and, when a battery is present in the circuit, converted to dc by means of a rectifier.

Figure 5-24 shows a vestigial charging circuit, used by Clinton, to power a headlamp or other accessories. A pair of charging coils mount on the ignition coil armature and receive power from 10 small flywheel magnets. Because the accessory load does not discriminate between ac and dc, no rectifier is necessary. Output can be checked with a PR 12 flashlight bulb connected across the "hot" and ground wires. Remove the spark plug and crank the engine. The system works if the bulb glows.

Figure 5-25 illustrates the next stage of development. The same paired lighting coils and flywheel magnets are present, but output passes through a selenium rectifier, which converts ac to dc by imposing high resistance on current flow in one direction. Part of the rectified output goes to the accessories and the remainder to replenish the battery.

One interesting feature of the circuit is the selenium rectifier (shown between the terminal block and the alternator). Today, almost all charging systems rectify by means of one or more silicon diodes that, while mechanically fragile and hypersensitive to polarity reversal, are inexpensive. To check the selenium type, tag and disconnect the four wires going to it. Connect the ohmmeter or continuity lamp leads between a center and side terminal. Note the meter response and reverse leads. The associated rectifier plate should conduct in one direction and resist current flow in the other. Repeat the test for the remaining pair of terminals. Check the alternator output voltage between both of the ac terminals and ground with the B wire disconnected.

Figure 5-26 shows a variation on the theme, this time a fuse, a heavy-duty key switch, and a silicon diodes are used. Check the fuse and fuse holder for continuity and test-system voltage output. Disconnect the battery wire at the B+ rectifier panel terminal and connect an alternator between this terminal and ground. Because the system is unregulated, voltage increases with engine rpm. This particular Tecumseh 3A circuit should deliver at least 12V at 2,500 rpm, 16V at 3,300, and 18V at maximum governed speed of 3,800 rpm. Note that not all unregulated charging circuits are designed to operate at these relatively high voltages and can be functioning normally when max output is on the order of 11.5V.

Check diodes for this and other systems with ohmmeter or

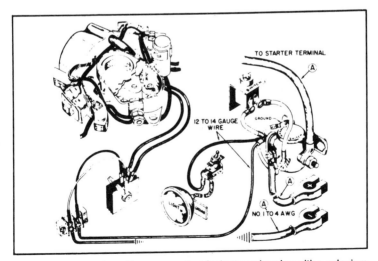

Fig. 5-25. Clinton adapts the same system for battery charging with a selenium rectifier in series with alternator output.

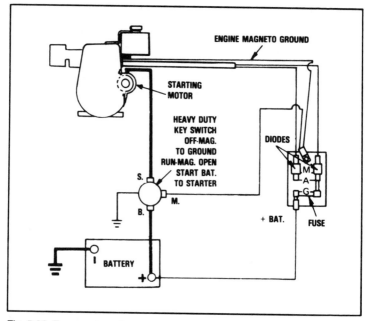

Fig. 5-26. Tecumseh and most other manufacturers employ silicon diodes for rectification. A manual switch engages the starter motor and a fuse protects stator windings from excessive current draw.

continuity lamp. A good diode conducts in one direction only, a shorted diodes conducts in both directions (usually ruining the battery in the process), and an open diode conducts in neither direction.

At this point I have discussed alternator-only and alternator-with-rectifier circuits. These circuits, rated between 14 and 30A, employ a solid-state regulator-rectifier encapsulated in epoxy or housed in a finned aluminum box (Fig. 5-27). In the standard format, three wires go to the unit: the two outboard wires from the alternator stator (sometimes marked ac) and the center, or B, wire to the battery. The regulator-rectifier is grounded through one or more hold-down screws.

Note: Check wiring arrangement because older and foreign-made units can vary from the standard format.

Solid-state electronics have become more reliable in recent years, but remain vulnerable to damage during service operations. Observe these prohibitions:

● Do not reverse battery polarity. Only a small minority of

regulator-rectifiers include a blocking diode to protect against this mishap.

● Make certain regulator-rectifier hold-downs are securely ground to engine (if necessary with a jumper wire).

● Disconnect the wiring harness at the regulator-rectifier before arc welding on engine-grounded equipment.

● Do not short alternator stator leads together or to ground.

● Do not disconnect the battery while the engine is running.

The previous point needs amplification. While there are exceptions, most regulator-rectifiers will suffer irreversible damage if the battery—that serves as a voltage-limiter—is removed from the circuit. This means that the open-circuit B+ output test, described for the charging circuits previously discussed, cannot be used. However, ac leads from the alternator can be disconnected at the regulator-rectifier for voltage checks.

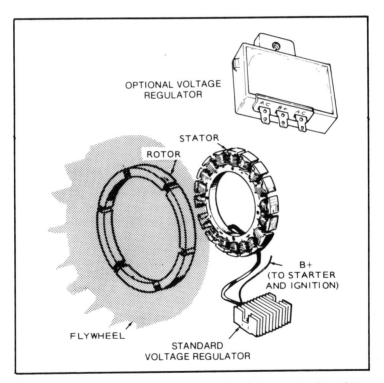

Fig. 5-27. Onan charging systems may use plastic encapsulated regulator-rectifier or familiar aluminum unit, heavily finned for cooling. In either case, ac leads are outboard with B+ tap in the center.

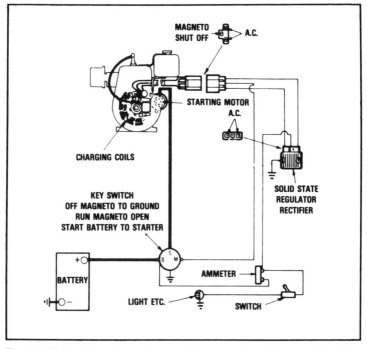

Fig. 5-28. Typical charging system used in conjunction with magneto ignition on Tecumseh engines. Note that the center wire in the engine connector is the magneto ground.

Figures 5-28 and 5-29 are reasonably typical wiring diagrams for magneto and battery-and-coil sparked engines. Battery and accessories feed from the B+ regulator-rectifier terminal in all cases.

Test values vary with make and model (although a general description remains possible). The primary situation is no charge or insufficient battery charge. If necessary, further discharge the battery with a headlamp or other load to drop terminal voltage to 12.5V. Start the engine with the same load connected and run for a few moments at 2500, 3000, and 3600 rpm. If voltage across battery terminals increases, the system can be declared functional. If voltage does not increase, connect the voltmeter across the ac leads to the regulator-rectifier as shown in Fig. 5-30. Note that the regulator-rectifier remains connected to both alternator and battery. Run the engine at full governed speed and observe voltage readings. Most alternators are good if about twice rated voltage—23 to 27V— develops between stator leads. If voltage is lower the stator is faulty.

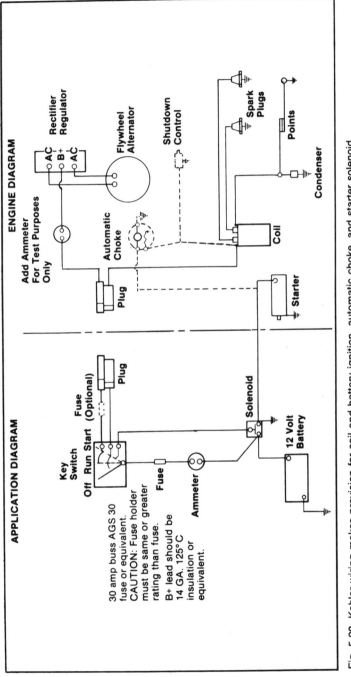

Fig. 5-29. Kohler wiring makes provision for coil and battery ignition, automatic choke, and starter solenoid.

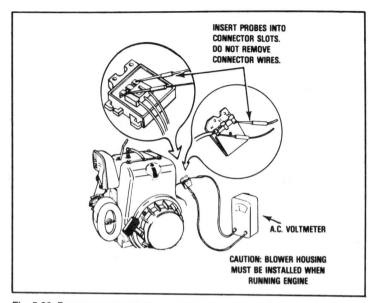

INSERT PROBES INTO
CONNECTOR SLOTS.
DO NOT REMOVE
CONNECTOR WIRES.

A.C. VOLTMETER

CAUTION: BLOWER HOUSING
MUST BE INSTALLED WHEN
RUNNING ENGINE

Fig. 5-30. Because most regulator-rectifiers do not tolerate battery disconnects and because ac tests described here require that the regulator-rectifier remain in circuit, test leads must be piggybacked to terminals. Insert test probes into connector slots without disconnecting the associated wiring. This regulator-rectifier mounts inside the blower housing that must be removed for access. Blower housing is reinstalled before startup and regulator-rectifier should be grounded at hold-down bolt.

If voltage is higher, the regulator-rectifier should be replaced.

Chronic overcharge can be detected by checking voltage output at full governed rpm across the battery terminals. Output should not exceed 14.7V and may comfortably be lower. Check ac voltage for a high reading that would point to a defective regulator-rectifier.

Chapter 6

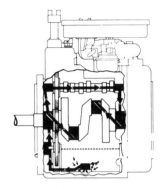

Engine Mechanics

Serious engine work involves some appreciation of mechanical theory, a number of specialized skills, and complete factory specifications for the engine at hand. Theory is necessary in order to understand why a component has failed. It is not enough to merely replace a broken connecting rod. You should be able to identify and correct the underlying cause of failure that, in this case, might involve the lubrication system, the governor mechanism, or connecting rod bolt locks.

The skills required include how to install a piston without breaking the rings, how to use reamers and other cutting tools and how to fit bearings. The British term "engine fitter" is appropriate because most engine work comes down to fitting parts to very precise clearances. And while this precision can be generalized about—for example, most small engines require 0.0015 of an inch running clearance between the crankpin and connecting rod bearing—there is no substitute for detailed, complete, and current factory specifications. If, for some reason, you cannot obtain a factory manual for the engine, at least obtain the specifications, available from a dealer and compiled in sources such as *Chilton's Small Engine Repair*.

But books are not enough. The mechanic must, first of all, come to terms with the engine in question. Study its architecture before disassembly and during the repair process. Scrupulously cleaning parts, lingering over them with brush and kerosene, is perhaps more beneficial to the mechanic than to the parts. You need the time to get the feel of the engine to "think," as Charles Kettering said, "like a piston." This kind of intuitive thought combined with some engineering theory, is what being a mechanic is about. We tell people we repair engines, but really it is a kind of high play only incidentally related to the hardware.

While the real work of engine repair centers on accurate fitting, beginning mechanics are sometimes daunted by the magnitude of the assembly process. Even the relatively simple utility and industrial engines discussed consists of hundred parts. There are, fortunately, some ways around the confusion.

Assembly is, nearly always, the mirror image of disassembly. What you take off first goes on last. Work on a clean, well-lighted bench and lay out the parts as they are removed from the engine. If necessary, make sketches of complex assemblies and, as circumstances permit, work in stages. Refurbish the cylinder head before the crankshaft is extracted. Do not disassemble the carburetor until the engine is buttoned up, and so on. Nuts, bolts, and washers should be positioned next to the components they secure. In addition, make a running notation of work to be done and replacement parts to be purchased.

DIAGNOSIS

Turn the engine over a few revolutions to detect possible binds and to establish that the connecting rod has not parted company with the piston. Excessive drag as the crankshaft is turned can have many causes, but is usually associated with a bent crankshaft or a severely galled cylinder bore. The classic symptom of a thrown rod is effortless rotation to about mid-stroke; then progress stops with a dull clunk. In severe cases, the crankshaft will have driven the parted rod end through the side of the block.

Some notion of connecting rod bearing wear can be had by gently rocking the flywheel a few degrees on each side of top dead center. This determination is made at tdc because crankpin/connecting rod bearings are subject to greatest wear at this part of the stroke and because engine geometry is favorable. During this part of the stroke, relatively large angular flywheel displacements are needed to absorb bearing clearances. Increased drag as the flywheel moves off of tdc signals that bearing clearance is taken up and that the piston is moving. In general, a new engine will have 3° or 4° of rock between piston movements. Wear may be considered excessive if the flywheel can be moved 10° between piston engagements.

Move the flywheel axially (as if you were attempting to pull it out by the roots). Main bearing endplay should be about 0.004 inch, or enough to make a "click" as the crankshaft shoulders against the thrust bearing. Excessive endplay is not, in itself, serious but it can indicate a generally high level of bearing wear.

Grasp the flywheel with both hands and attempt to push it from side to side. Radial clearance should be minimal—on the order of 0.02 inch. Much more side play than this means severe wear on main bearings and possibly the crankshaft. Either or both crankcase seals may be affected, resulting in oil leaks (four-cycle engines) or air leaks (two-cycles).

Next check crankshaft straightness. A rough and ready way to do this is to visually track the crankshaft centering holes—chamfered holes drilled in both ends of the crankshaft—while the flywheel is spun. Remove the spark plug(s) and check the pto end first; it is more likely to be bent. The centering hole should track in a true circle without perceptible wobble. Repeat the operation on the flywheel end. Loose mounting bolts and enlarged engine shroud bolt holes are secondary indications of crankshaft problems.

Mechanics traditionally make a compression test before embarking on major repairs. However, many small engines, including Briggs & Stratton and Kohler singles (except K 91), incorporate a compression release on the exhaust valve. The Briggs unit is engaged by the starter and the Kohler design employs centrifugal weights to keep the valve from making a gas-tight seal below 600 rpm. In either case, the engine must be spun backwards for meaningful compression readings. Kohler suggests that a gauge pressure of 90 to 100 psi is normal after seven or eight successive compression strokes. Briggs & Stratton gives no compression specification, but says that the flywheel should rebound when spun backwards with the spark plug in place.

Engines without automatic compression release mechanisms can be tested as follows:

● Remove spark plug and ground ignition.
● Install a compression gauge in cylinder-head spark plug boss. A 14mm threaded gauge adapter is more convenient than the friction type (Fig. 6-1).
● Open throttle and choke full wide.

Fig. 6-1. A compression test is a fairly reliable barometer of engine condition, but best results can be had with a compression history made once every six months or so as part of routine maintenance.

- Spin flywheel at normal starting speed.
- Record gauge reading on the seventh compression stroke.

At this point, preliminary inspection is complete and, whenever possible, the engine should be started for more detailed diagnosis. Figure 6-2 and 6-3 show causes of excessive oil consumption and loss of power in four-cycle engines. As far as two-cycles are concerned, excessive oil consumption (that can contribute to loss of power and overheating) can only be caused by an improper fuel mixture. Loss of power in two-cycle engines is almost always traceable to worn piston rings or worn piston and cylinder bore. Note that while two-cycle engines have excellent oiling systems that assures a fresh lubricant with each crankshaft revolution, piston and rings are heavily stressed. These engines fire every revolution and compress the mixture at both ends of the cylinder bore. Consequently, ring and bore life is somewhat shorter than for equivalent four-cycle designs.

Engine knocks are usually "spark knocks" that on both engine types can be traced to improper ignition timing or excessive carbon

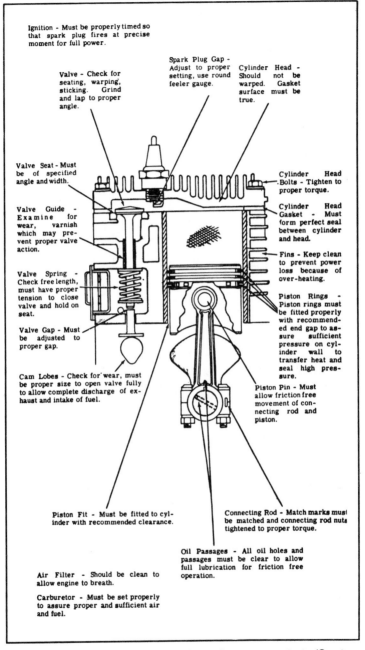

Ignition - Must be properly timed so that spark plug fires at precise moment for full power.

Spark Plug Gap - Adjust to proper setting, use round feeler gauge.

Valve - Check for seating, warping, sticking. Grind and lap to proper angle.

Cylinder Head - Should not be warped. Gasket surface must be true.

Valve Seat - Must be of specified angle and width.

Cylinder Head Bolts - Tighten to proper torque.

Valve Guide - Examine for wear, varnish which may prevent proper valve action.

Cylinder Head Gasket - Must form perfect seal between cylinder and head.

Fins - Keep clean to prevent power loss because of over-heating.

Valve Spring - Check free length, must have proper tension to close valve and hold on seat.

Piston Rings - Piston rings must be fitted properly with recommended end gap to assure sufficient pressure on cylinder wall to transfer heat and seal high pressure.

Valve Gap - Must be adjusted to proper gap.

Cam Lobes - Check for wear, must be proper size to open valve fully to allow complete discharge of exhaust and intake of fuel.

Piston Pin - Must allow friction free movement of connecting rod and piston.

Piston Fit - Must be fitted to cylinder with recommended clearance.

Connecting Rod - Match marks must be matched and connecting rod nuts tightened to proper torque.

Oil Passages - All oil holes and passages must be clear to allow full lubrication for friction free operation.

Air Filter - Should be clean to allow engine to breath.

Carburetor - Must be set properly to assure proper and sufficient air and fuel.

Fig. 6-2. Factors that affect four-cycle engine power output. (Courtesy Tecumseh.)

189

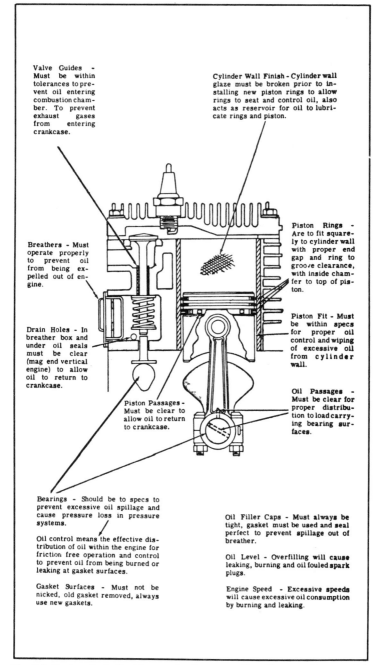

Valve Guides - Must be within tolerances to prevent oil entering combustion chamber. To prevent exhaust gases from entering crankcase.

Cylinder Wall Finish - Cylinder wall glaze must be broken prior to installing new piston rings to allow rings to seat and control oil, also acts as reservoir for oil to lubricate rings and piston.

Piston Rings - Are to fit squarely to cylinder wall with proper end gap and ring to groove clearance, with inside chamfer to top of piston.

Breathers - Must operate properly to prevent oil from being expelled out of engine.

Piston Fit - Must be within specs for proper oil control and wiping of excessive oil from cylinder wall.

Drain Holes - In breather box and under oil seals must be clear (mag end vertical engine) to allow oil to return to crankcase.

Piston Passages - Must be clear to allow oil to return to crankcase.

Oil Passages - Must be clear for proper distribution to load carrying bearing surfaces.

Bearings - Should be to specs to prevent excessive oil spillage and cause pressure loss in pressure systems.

Oil control means the effective distribution of oil within the engine for friction free operation and control to prevent oil from being burned or leaking at gasket surfaces.

Gasket Surfaces - Must not be nicked, old gasket removed, always use new gaskets.

Oil Filler Caps - Must always be tight, gasket must be used and seal perfect to prevent spillage out of breather.

Oil Level - Overfilling will cause leaking, burning and oil fouled spark plugs.

Engine Speed - Excessive speeds will cause excessive oil consumption by burning and leaking.

Fig. 6-3. Factors that affect four-cycle oil consumption. (Courtesy Tecumseh.)

accumulation in the combustion chamber. Mechanical knocks can have several sources, such a loose flywheel, a worn connecting rod, worn main bearings, or excessive piston clearance.

Seal leakage on two-stroke engines deserves special mention. Complete failure of either crankshaft seal means that the engine will not run because crankcase compression is required to pump fuel into the cylinder. An experienced mechanic can turn the flywheel by hand and feel crankcase pressure buildup. Partial seal failure results in loss of power and possible overheating. The tipoff is that the engine will demand very rich carburetor settings and might refuse to run unless the choke is partially or completely shut.

The best approach is to replace both seals at the first sign of trouble. Seal test tools—consisting of a squeeze bulb, pressure gauge and blanking plates for intake and exhaust ports—are available from some snowmobile dealers. To test, seal the engine block, pressurize to about 5 psi, and immerse in a tub of solvent. Seal leaks show as air bubbles around the crankshaft.

Two-cycle engines—particularly low-speed industrial types— can strangulate from carbon fouling at exhaust ports or the muffler. Some mufflers can be disassembled for cleaning. Steel (as opposed to aluminum) units can be soaked in a warm solution of household lye and water. To clean exhaust ports, remove the muffler or exhaust pipe, retract the piston below port level, and scrape with a screwdriver or dull knife (Fig. 6-3). Spin the engine a few times to clean carbon flakes from the cylinder bore. The spark plug might foul on first startup if loose carbon is present in the cylinder. See Fig. 6-4.

SCOPE OF WORK

There are three reasons to go into an engine. The first is to make a specific and limited repair, such as correcting a centrifugal governor problem or freeing a stuck exhaust valve. An *overhaul* is somewhat more generalized and consists of replacing seals, gaskets, rings, and, when detachable, connecting rod big-end bearings. Most shops also reseat the valves on four-cycle engines. A *rebuild* is the most ambitious procedure and, done correctly, involves restoration of every engine and accessory bearing surface to original specification. Worn parts that cannot be remachined are replaced. New factory paint and decals top off the job.

A limited repair is in order when the engine experiences a component failure early in its life. An overhaul could add 30 percent

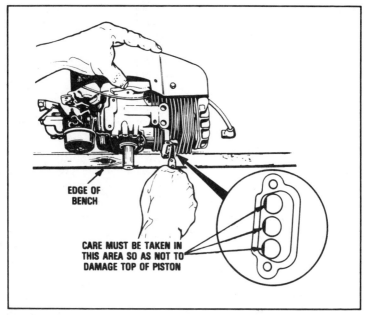

Fig. 6-4. Two-cycle exhaust ports should be periodically scrapped to remove carbon deposits. Tecumseh engine shown is typical in that the piston does not retract below port floors.

or so to the longevity of an otherwise sound powerplant, while a rebuilt engine is reserved for the worst cases (those that knock or pump clouds of oil smoke). The crankshaft usually decides the issue. If the crank is good, other bearing surfaces are probably tolerable, and you might be able to get by with an overhaul. If the crank is flat, bent, or otherwise unserviceable, the choice is between a new engine, a factory short block, or a rebuild of the existing unit. Short blocks are theoretically less expensive than complete engines and consist of a new block casting and all internal crankcase parts.

CYLINDER HEAD

Most utility and industrial engines employ demountable cylinder heads, sealed with throwaway composition gaskets and secured to the block by capscrews.

Warning: Composition gaskets employ asbestos as a filler. Dispose of the gasket in a safe manner. Carefully scrape gasket remains from the block and head without breathing or ingesting *any* dust that is generated.

With the engine at room temperature, remove the capscrews.

Note, variations in length such as are present on aluminum-block Briggs & Stratton engines. Some European makes employ studs that pass through or around the cylinder barrel and anchor in the crankcase. These same engines can be equipped with reusable copper head gaskets. Anneal the gasket by heating with a propane torch and quenching in oil or water.

Remove carbon deposits from the cylinder head, piston top, and block. An end-cutting wire brush is the preferred tool (Fig. 6-5), although a dull knife may also be used on stubborn deposits. Be careful not to gouge the aluminum or damage the gasket surfaces.

Inspect the spark plug boss for stripped or pulled threads. Repairs can be made with a 14mm Heli-Coil kit. Check head distortion with the aid of a surface plate or piece of plate (not window) glass. The head is considered acceptable if a 0.003-inch feeler gauge will not pass between bolt holes (Fig. 6-6). Ideally a warped head should be replaced together with the head bolts. If the engine is fitted with a single head casting, however, the head can be reground without serious side effects. Separate head castings, such as found on horizontally opposed twin-cylinder engines, may also be ground, but great care must be exercised to take off equal amounts of metal on both. Tape a piece of medium-

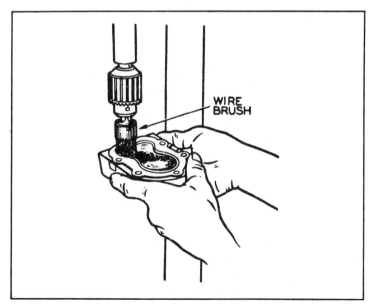

Fig. 6-5. Remove carbon deposits from cylinder head and engine block deck with an end-cutting wire brush.

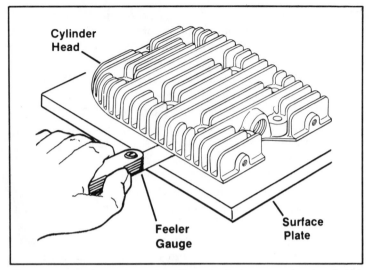

Cylinder Head

Feeler Gauge

Surface Plate

Fig. 6-6. Cylinder head flatness should be checked to assure head gasket integrity. Commercial plate glass can be substituted for the surface plate shown. (Courtesy Kohler.)

grit, wet-or-dry emory paper to the plate glass and, applying pressure at a point near the center of the head, grind the gasket surface. Oil speeds the process. When the gasket surface is uniformly bright, the head is flat.

Install a new gasket and torque in three equal increments to specification. Normally, cylinder head bolts are installed dry. Some mechanics lubricate threads with motor oil or assembly lube. When a lubricant is used, it is good practice to increase torque 10-15 percent over the factory limit. The torque sequence for four-bolt heads is a simple X pattern. Others are torqued from the center bolts outward so that the ends of the casting go down last. However, the factory might make exception to this general rule (as shown in Fig. 6-7). Consult the factory manual for the engine in question.

VALVES

Side valves, i.e., those located in the block, are removed and installed with either of the compressor tools shown in Fig. 6-8. Rotate the flywheel to seat the valve, insert the compressor under the valve collar, compress the valve spring, and remove valve locks. It is good practice to temporarily plug the oil drain hole in the valve chamber floor to prevent a valve lock from falling in the crankcase.

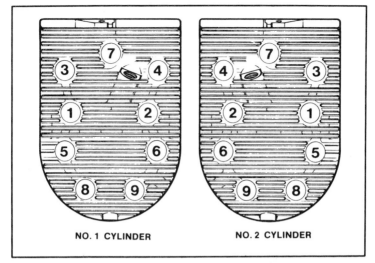

NO. 1 CYLINDER   NO. 2 CYLINDER

Fig. 6-7. Cylinder-head torque sequence varies with engine make and model. Onan B43E, 43G and 48G patterns differ between cylinders.

Split locks are almost universal; Tecumseh and, most notably, Briggs & Stratton use cross pins. Some Briggs models employ one-piece retainers. See Fig. 6-9. Installation is the reverse of disassembly. Split locks can be positioned with the aid of grease and a screwdriver as (shown in B of Fig. 6-8). Professional mechanics generally prefer to use a magnetic insertion tool such as Snap-on's CF 771.

When properly secured, split locks will be swallowed by the collar and will be no longer visible. Briggs' one-piece retainer will be centered under the collar and the cross pin will be tucked out of sight under the collar.

A side-valve engine can be serviced without special tools (although the procedure costs something in frustration). Lift the collar with two flat screwdrivers. The trick is to keep the collar level so the valve remains seated as the spring compresses. Split locks should fall or can be knocked free; other types are removed with long-nosed pliers. Installation is a bit more difficult, and particularly so when you are working alone. Compress the spring, as before, and make certain that the valve is seated. Hold pressure on the screwdrivers with one hand and insert the valve locks. Split keepers might slip partly out of the valve stem groove, but can be tapped into place against spring tension.

Overhead valve mechanisms are more accessible and,

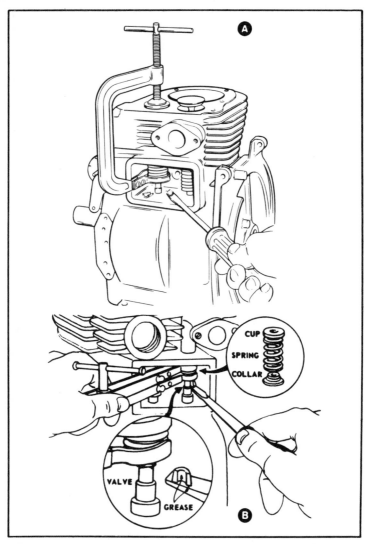

Fig. 6-8. Either a clamp (A) or bridge-type (B) compressor can be used to remove and install block-mounted valves. The former tool is available from Kohler, the latter from Briggs & Stratton. Note how split valve locks are spooned into place with the aid of a grease-coated screwdriver.

consequently, easier to service than side valve units. Detach the cylinder head and support the head and valves on a wood block sized to fit the combustion cavity. The purpose of the block is to prevent valve movement as springs are compressed. Collapse springs with a crow's foot tool (Fig. 6-10). An alternate disassembly

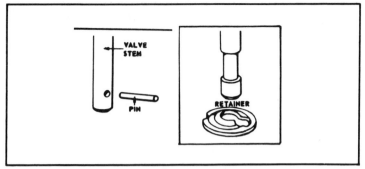

Fig. 6-9. In years past, Briggs & Stratton used a pin-type valve lock. Some later production employs a slotted retainer (also used by Tecumseh).

technique is to position a large wrench socket over the collar and rap the socket sharply with a hammer. Impact will simultaneously compress the valve spring and dislodge split keepers. Assemble with a crow's foot tool.

Valve springs are often, but not always, interchangeable between intake and exhaust sides. When springs differ, the heavier or, as the case may be, double, spring serves the exhaust valve. Some engines employ springs with closely wound damper coils that should be assembled on the stationary end of the spring (Fig. 6-11).

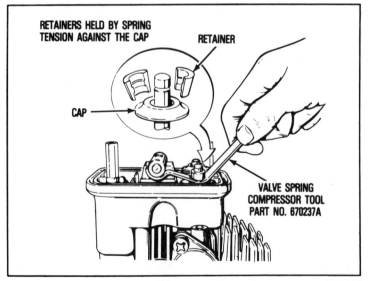

Fig. 6-10. Tecumseh overhead valve gear, showing split-type valve locks and factory spring compressor.

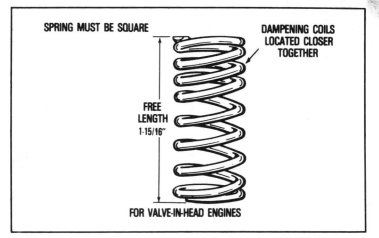

SPRING MUST BE SQUARE

DAMPENING COILS
LOCATED CLOSER
TOGETHER

FREE
LENGTH
1-15/16"

FOR VALVE-IN-HEAD ENGINES

Fig. 6-11. Damper coils on Tecumseh ohv and other engines should be on the stationary end of the spring, away from the valve locks.

In other words, the tightly wound coils are positioned farthest away from the rocket arm or valve lifter.

Valve springs are replaced as a matter of course during an engine rebuild and should be replaced during a conscientious overhaul. Weak springs can cause a hard-to-diagnose high-speed miss and, on ohv designs, can become detached with disastrous consequences. Nevertheless, original springs can be reused if:

● The spring stands flat.
● The free-standing height meets the manufacturer's specification.
● There is no evident of stress pitting or contact between adjacent coils.

Valve nomenclature is illustrated in Fig. 6-12. Inspect intake and exhaust valves, which incidentally are not interchangeable, for deep pitting, cracks and stem distortion. Leaks across the valve face cost compression, a condition signaled by hard starting, loss of power and, on the intake side, by "pop back" through the carburetor.

Figure 6-13 illustrates normal and abnormal valve wear patterns. Valve A has run almost 1000 hours in a Kohler test engine and, despite the heavy accumulation of combustion products on heat and upper stem, can probably be reconditioned. Valve B suffers from severe stem corrosion, caused by moisture in fuel or from condensation. The latter condition occurs when the engine is improperly preserved during extended layup or when engine is repeatedly stopped before normal operating temperatures develop. The valve should be replaced.

Valve C is badly worn. It is the victim of too many hours between overhauls. Margin has become knife-edged and the head is hopelessly warped. Valve D shows evidence of chronic leakdown. Because the face is gas cut, the valve cannot be reconditioned and must be replaced. Valve E shows coking, or carbon buildup, that is a normal and relatively benign condition for intake valves. Clean and regrind. Gum deposits, valve F, usually result from stale gasoline and are often seen when fuel is left in the tank during engine storage. Generally, the valve and guide can be cleaned and reused but in severe cases the valve must be replaced.

An overheated exhaust valve, G, is frequently encountered. Note the dark discoloration of the upper stem and absence of combustion deposits. The valve should be renewed and cause of the problem should be corrected. Look for maladjusted ignition timing, overly lean carburetor adjustment, blocked cooling fins, a weak valve spring or worn valve guides. Carbon cut valve, H, has been destroyed by carbon buildup in the combustion chamber. This condition can been avoided with proper maintenance. Carbon accumulates rapidly in engines that run at part throttle under light load.

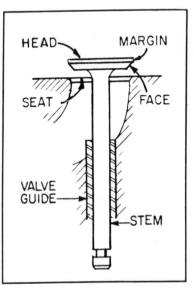

Fig. 6-12. Basic valve nomenclature. (Courtesy Clinton.)

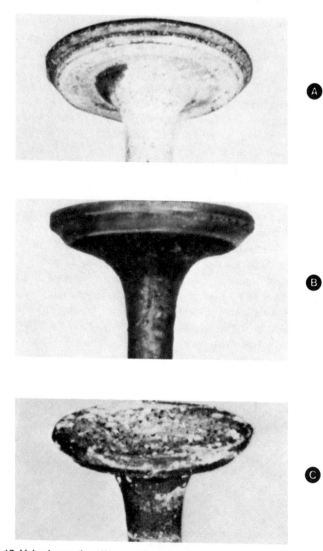

Fig. 6-13. Valve inspection: (A) normal wear and deposit accumulation on valve that ran almost 1000 hours under laboratory conditions; (B) severe stem corrosion from water in fuel or condensation (replace valve because stem pits act as stress risers); (C) extreme wear, characterized by thin margin and warped head mean that valve should be replaced.

Worn valve faces and seats should be turned over to a dealer or competent automotive machinist for servicing. The cost of the tools makes this work prohibitive for the casual mechanic. See Fig. 6-14.

Fig. 6-13. Valve inspection: (D) gas channeling at valve face, probably caused by an improper valve grind; (E) heavy carbon buildup on intake valve, normal flow slow-turning engines, (valve usually can be reused after cleaning); (F) gum deposits from stale gasoline, one of the major causes of valve sticking (ream guides and clean valves).

Figure 6-15 shows a commercial valve grinder in use. While most small engine valves are cut at 45°, Onan likes 44°. Some Briggs & Stratton models have 30° intake valves and 45° exhausts. The moral of all this is that valve work requires factory documen-

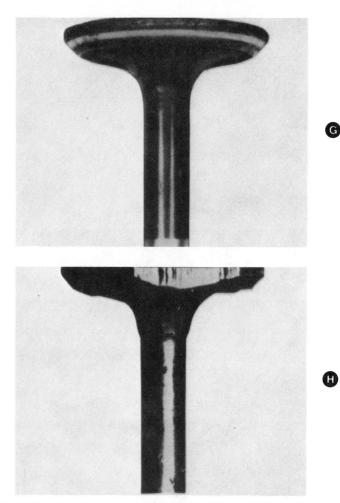

Fig. 6-13. (G) overhead exhaust valve with dark discoloration above guide. (Problem may be caused by worn guides or age-weakened springs. Also check ignition timing, carburetor adjustment, and cooling system. Valve should be replaced); (H) carbon-cut valve destroyed by carbon deposits in head cavity. (Courtesy Kohler.)

tation for the particular make and model. At any rate, the valve face is cut at a single angle and should leave a margin (see Fig. 6-12) of about 1/64 of an inch. Less than this will cause the valve to overheat and might send the engine into detonation.

The shop should also be able to handle seat refinishing (although automotive machinists do not always have the appropriate pilot). Normally, a high-speed grinder is used, but cast-iron, or

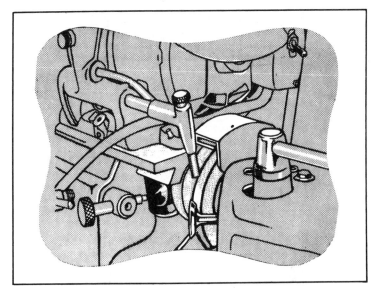

Fig. 6-14. A valve lathe is used to reface valves. Grinder may also be used to widen margin at some cost in valve diameter. (Courtesy Clinton.)

Fig. 6-15. Seats can be refurbished by hand or with a portable grinder and suitable "rocks." Seat width, angle and entry angle are crucial.

integral, seats can be refurbished with a relatively inexpensive reamer of the type shown in Fig. 6-16. Valve seat angle and seat width are matters of specification, but the angle is always 1/2 to 1° larger or smaller than the valve face angle in order to provide an interference fit. Seat width is controlled by entry and exit angles (as illustrated in Fig. 6-16). An overly narrow seat soon hammers out under valve impact and a seat that is too wide makes a poor seal.

While there is always some loss of control when work is farmed out, you can expect the machinist to accurately reproduce original valve dimensions, unless some metal gets lost in the grinding. Provide the specifications, if he does not have them, and ask that grinding be kept to a minimum. "Buried" valves have poor flow characteristics and reduce available spring tension. Springs, however, can be shimmed at their stationary ends with hardened washers available from bearing supply houses.

Most manufacturers suggest that valves should be lightly lapped after machine refinishing. Few professional mechanics take the time to do this, but lapping does ensure a perfect seal.

Obtain a suction cup tool, sized for small engine work such as K-D Tools' catalog No. 501, and tin of Clover Leaf oil-based, valve-grinding compound. Dab a small amount of compound on the valve face, insert the valve and mount the suction cup tool as shown in Fig. 6-17. The cup might not find purchase on highly polished valve heads and some form of adhesive can be used. Rotate the tool between your palms, stopping when the compound degrades and no longer makes the characteristic hiss as the valve is worked. Raise the valve from its seat, spot a little more compound around the face, rotate the valve a quarter-turn, and repeat the operation. Stop when the valve face and seat take on a uniform met finish. Wipe all traces of compound from the valve, seat the valve chamber, and flush with solvent. Compound that remains in the engine will continue its work on valve stems and guides.

At the risk of repetition, it should be pointed out that valve lapping is a touch-up operation that should not require more than 30 seconds per valve. Excessive lapping will groove the valves and produce the condition shown in Fig. 6-18. When the engine is cold, the groove and seat match perfectly. Once the engine attains operating temperature, the valve expands away from the seat and leaks.

Valve Guides. Integrity of the valve seat depends, in large measure, upon the condition of the valve guide that centers the valve (and reconditioning tool) on the seat. It is a waste of time

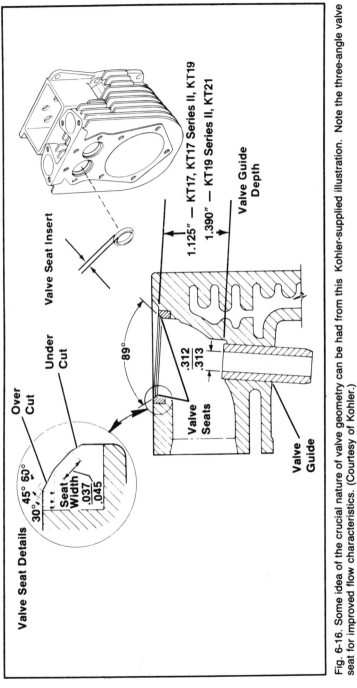

Valve Seat Details

Over Cut

Under Cut

45° 60°
30°

Seat Width
$\frac{.037}{.045}$

Valve Seat Insert

89°

Valve Seats

$\frac{.312}{.313}$

1.125" — KT17, KT17 Series II, KT19
1.390" — KT19 Series II, KT21

Valve Guide Depth

Valve Guide

Fig. 6-16. Some idea of the crucial nature of valve geometry can be had from this Kohler-supplied illustration. Note the three-angle valve seat for improved flow characteristics. (Courtesy of Kohler.)

to grind a valve that rides in a sloppy guide.

Guide-to-valve-stem clearance is measured at the top (valve head end) of the guide, and any figure of 0.0045 inch or more is excessive. Minimum clearance is quite small (on the order of 0.0015 inch). Briggs & Stratton supplies their dealers with plug gauge sets to make these determinations. Experienced mechanics check valve guide wear as a function of how much the valve wobbles when full open.

Most engines have some form of replaceable guide that is pressed into place. A few of the least expensive models run the intake and sometimes the exhaust valve directly against block metal, but there is always provision to retrofit guides. Many Briggs & Stratton guides are repaired by partial reaming and installation of a bushing in the upper guide area. And most manufacturers supply valves with oversized stems that can be installed after the original guides are reamed to fit.

If you establish that guides are worn, the first step is to obtain the necessary replacement parts and reamer. The later can be purchased from a good tool supply house and the new guides, guide

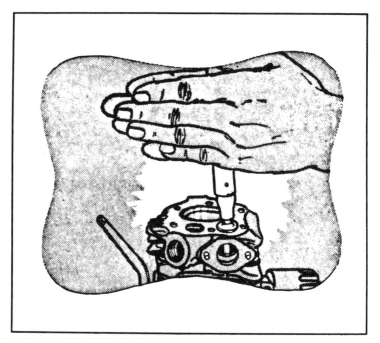

Fig. 6-17. Lapping assures a positive seal after valves and seat have been reworked.

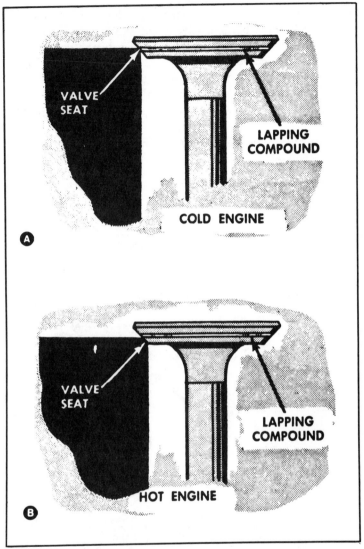

Fig. 6-18. Lapping is no panecea and will not substitute for machining operations. When an engine is cold, a lapped seat will work. As the engine reaches running temperature, the valve expands away from lapped surface and leaks.

bushings, or oversized valves should be obtainable from a dealer (together with detailed installation instructions).

Replaceable guides are knocked out from the valve head end after measuring the distance between the top of the guide and the valve seat insert. Side valve guides sometimes have to be broken before extraction from the valve chamber. Others can be withdrawn in one piece. Clean parts in solvent and—working from the valve head end—drive replacement guides home to the depth of the originals. Valve guide drivers that center on the guide ID can be purchased from specialty tool houses. Use of the appropriate driver may eliminate the need to finish ream the guide to 0.0015-0.002 inch larger than the valve stem. Otherwise, the guide must be reamed.

Figure 6-19 shows how Briggs & Stratton valve guide bushings are superimposed upon the original guides. This job requires special

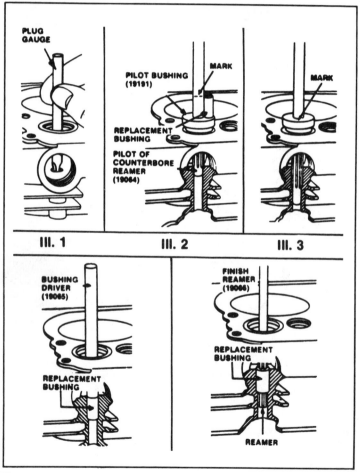

Fig. 6-19. Installation of valve guide bushings in Briggs & Stratton engines is a job best left to a dealer.

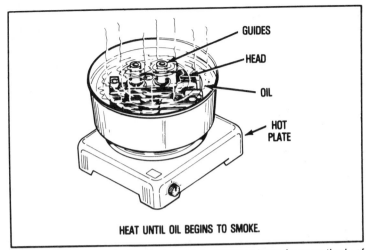

HEAT UNTIL OIL BEGINS TO SMOKE.

Fig. 6-20. Aluminum ohv heads do not take kindly to brute-force methods of valve guide extraction and installation. The head should be heated and replacement guides chilled.

tools and should be farmed out to a dealer. Figures 6-20 and 6-21 depict another approach to guide service. In this example, guides are mounted in an aluminum head. The head is immersed in oil and heated to 375-400° F—a process that should be done

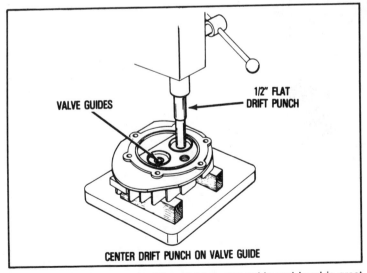

CENTER DRIFT PUNCH ON VALVE GUIDE

Fig. 6-21. If the temperature differential between guide and head is great enough, guide would drop into place. As a practical matter, an arbor press is needed.

outdoors—and the old guides are pressed out. Then, the head is brought back up to temperature and new guides, which have been chilled, are pressed in. Because installation depends upon thermal expansion and contraction, little violence is done to the guides. Finish reaming should not be necessary.

Note: Valves and valve seats must be ground and lapped after guide installation.

Valve Seats.. Most engines feature replaceable valve seats that must be renewed in event of looseness, cracking, or deep pitting (a seat insert is shown back in Fig. 6-16). Briggs & Stratton cast iron block models employ a replaceable exhaust valve seat, but the intake valve runs directly on the block. Briggs is nothing if not thorough, however, and intake valve seat inserts can be installed with proper tools as described as follows.

Valve seat replacement is a relatively unusual service operation and parts should be obtained before embarking upon such work. Remove the old seat with a long punch (Fig. 6-22) or, if that is not possible, with a purchased or homemade removal tool (Fig. 6-23).

Most replacement valve seats—particularly those intended for installation in cast-iron blocks—have the same OD as the original seat. This eliminates the need for a special reamer, but limits the

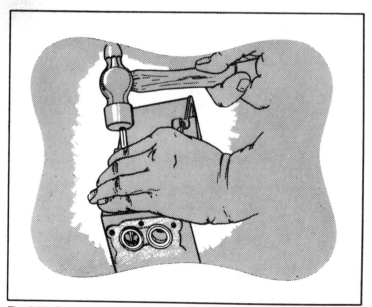

Fig. 6-22. Sometimes it is possible to drive out valve seats from below, as in this Clinton block.

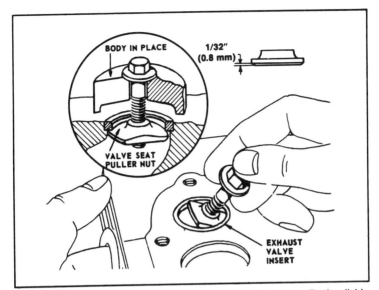

Fig. 6-23. When ports are restrictive, a hammer impact bushing puller (available from auto parts houses) or Briggs & Stratton's part No. 19138 can be used.

repairability of the block when the seat has loosened and wallowed its recess. The wear limit for Briggs & Stratton engines is 0.005 of an inch clearance between the seat OD and block ID (Fig. 6-24). The alternative approach is typlified by Clinton in their GEM series engines. Replacement seats are 0.040-inch oversized and a reamer, piloted on the valved guide, must be used to enlarge the recess.

As mentioned earlier, intake valve seats on Briggs & Stratton cast iron engines must be reamed to accept an insert. The factory provides its dealers with a well-engineered tool for this purpose (Fig. 6-25).

Fig. 6-24. Loose valve seats are not unknown in aluminum block engines. If a seat with an oversized OD is not available, the problem can usually be corrected by staking the seat.

211

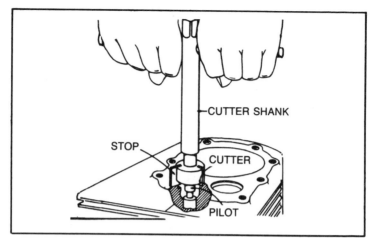

Fig. 6-25. A scrap valve can substitute for the factory valve seat driver shown.

Replacement seat and recess must be dry and spotlessly clean. Chill the seat and, working quickly, press it into place. A factory driver, such as the one shown in Fig. 6-26, should be used when available. Otherwise use a scrap valve. It is good practice, especially on aluminum block engines, to stake the newly installed seat. Begin

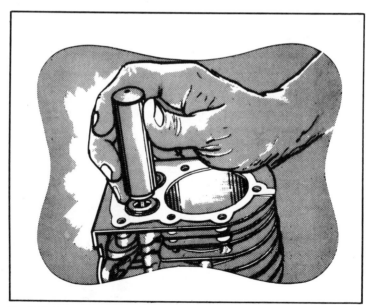

Fig. 6-26. Valve seat reamer in use. Note that this tool pilots on the valve guide and will not cut true if the guide is worn.

at three points 120° apart and complete the job at close intervals around the seat circumference.

The associated valve should be replaced or, if still serviceable, reground and lapped to the seat.

Valve Lash Adjustment. Metal lost through grinding and lapping operations and metal gained through seat or valve replacement must be compensated for by valve lash adjustment. See Fig. 6-27. Side valve engines generally are fitted with fixed tappets and valve lash increases are made by grinding the valve stems. Install the valve without spring, turn the crankshaft until the tappet fully retracts and measure the clearance between the tappet and valve stem with a feeler gauge (Fig. 6-28). Grind the valve stem as necessary to establish factory-specified clearance (on the order of 0.008 inch intake and 0.010 inch exhaust). Work slowly,

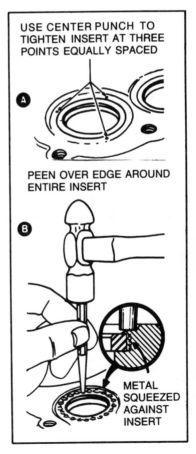

Fig. 6-27. Staking the valve is a good and perhaps necessary practice when replacing seats in aluminum. First nail the seat down with three stakes about 120° apart (A). Secure with stakes around its whole perimeter (B).

USE CENTER PUNCH TO TIGHTEN INSERT AT THREE POINTS EQUALLY SPACED

PEEN OVER EDGE AROUND ENTIRE INSERT

METAL SQUEEZED AGAINST INSERT

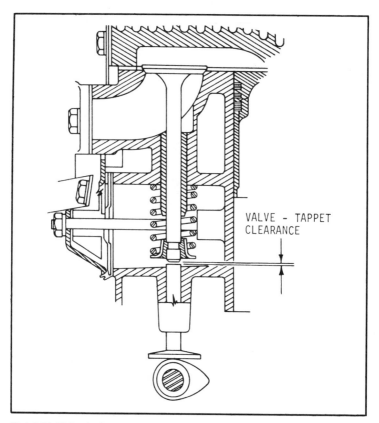

Fig. 6-28. Valve lash measured between stem end and tappet with tappet on heel of cam. Kohler engine shown closely resembles other side valve types.

frequently rechecking the lash, and exercising care that the stem remains dead flat. If too much metal is lost, the lash will be excessive and valve timing will retard for some loss of power. Correct by regrinding and lapping the valve face.

Some of the better side-valve engines and all overhead valve types have adjustable tappets. Valve lash for ohv engines is defined as clearance between the rocker arm and valve stem (Fig. 6-29). Turn the crankshaft until the associated tappet is on the heel of its cam lobe, loosen the lock nut, and turn the adjustment nut to achieve specified clearance. Tighten the lock nut and recheck.

Valve Gear Modification. It is sometimes possible to upgrade standard-duty engines to heavy-duty status by parts substitution. Table 6-1 gives the interchange parts numbers for Briggs & Stratton engines.

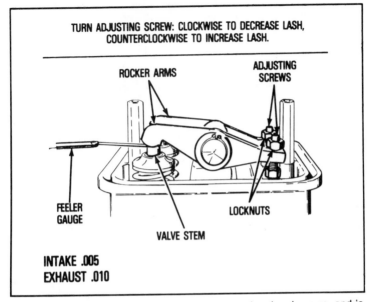

ROCKER ARMS

ADJUSTING
SCREWS

FEELER
GAUGE

LOCKNUTS

VALVE STEM

INTAKE .005
EXHAUST .010

Fig. 6-29. OHV lash is measured between stem end and rocker arm, and is adjusted via threated push rod pivots.

Breather. Four-cycle engines incorporate some form of breather assembly (Fig. 6-32). The breather consists of a check valve, bleed port and oil trap. The check valve opens on the piston downstroke—when crankcase pressures are highest—to allow air to be expelled from the crankcase. It remains closed for the rest of the stroke, sealing the crankcase and keeping its pressure at some value below atmospheric.

Partial vacuum tends to reduce oil seepage at gaskets and crankshaft seals. However, the breather bleed port remains open during the whole cycle. Consequently, some air enters the crankcase to be circulated and expelled, together with combustion gases, on the next downstroke. The oil trap prevents escape of lube oil out of the breather tube. Exhaust may be vented to the atmosphere or to the carburetor intake.

Some two-cycle engines use a functionally similar device for an entirely different purpose. The Reed assembly (Fig. 6-31) acts as a check valve to contain the air-fuel mixture in the crankcase.

PISTONS AND RINGS

Engine configuration determines piston access. Single-cylinder utility and industrial plants typically employ unitized have cylin-

Table 6-1. Briggs & Stratton Stallite Valve and Torocap Conversion.

	Stellite valve	Rotocap only conversion			
		Spring	Rotocap	Retainer	Pin
Aluminum					
60000, 80000, 82000, 92000, 94000	260443	26826	292259	230127	230126
100000, 130000	260860	26826	292259	230127	230126
140000, 170000, 190000, 200000, 250000	390420	26828	292260	93630	
Cast Iron					
14, 19, 190000, 200000	26735	26828	292260	68283	
23, 230000	261207	26828	292260	68283	
240000, 300000, 320000	261207	26828	292260	68283	(Stellite Std.)

216

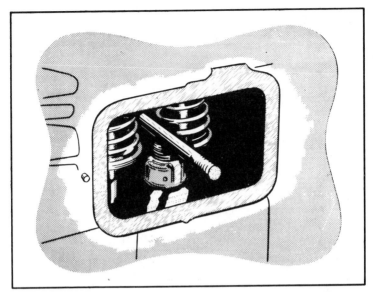

Fig. 6-30. Conical crankcase breatherb with filter and check valve used in some Clinton engines. Other types are integral with valve chamber cover or may be remotely mounted and connected to crankcase by a hose.

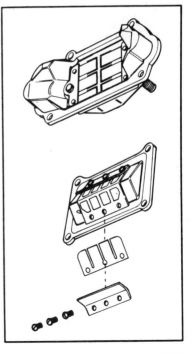

Fig. 6-31. Two reed valve assemblies, flat (upper) and dished (lower) for increased flow. Dished unit employs stop plate to limit reed deflection.

der barrels that are cast in one piece with the block. The head is detachable and the connecting rod is split at the crankpin. One side of the crankcase is closed with a cover, whose arrangement depends upon the lay of the crankshaft:

Horizontal Crankshaft—cover on power takeoff side of block, top-side main bearing or cover, which serves as oil pain, on bottom of block.

Vertical Crankshaft—cover (known as *flange*) on bottom of block, supports lower main bearing and engine mounting bolts.

Figure 6-32 shows a Kohler horizontal-shaft block being lifted off its combination crankcase cover-oil pan. No problems here. All you have to remember is to renew the gasket upon assembly. Flange or side-cover engines are a little more complicated (Fig. 6-33). Polish all traces of rust or paint from the crankshaft extension. Use strip emory cloth and remove burrs and break sharp edges on keyways with a fine file. Take your time with this operation, dressing out all imperfections on the crankshaft that would bind the main bearing or cut the lower oil seal. Mount the engine upright, lubricate the crankshaft, remove the flange hold-down bolts and, using a soft mallet, gently tap the flange away from the block. Carefully

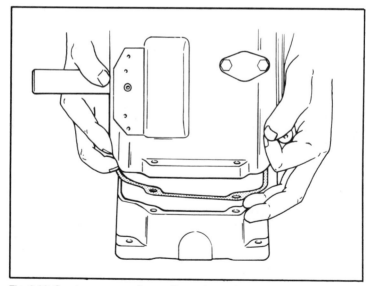

Fig. 6-32. Crankcase cover-oil pan offer easiest rod access because main bearings are not disturbed.

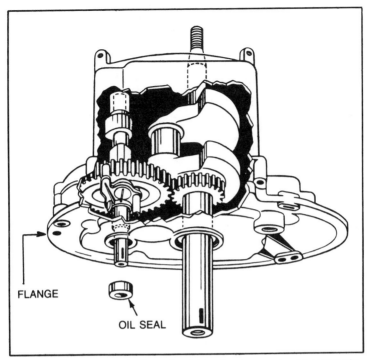

Fig. 6-33. Flange on vertical crank engines locates crank and camshafts. Remove with extreme care, cleaning shaft extensions and making the initial parts separation with a rubber hammer. Do not use a screwdriver to jimmy the flange free of the block.

withdraw the flange. The camshaft should remain in mesh with the crankshaft.

The connecting rod is secured with two bolts or studs that can be proofed against vibration loosening by lockwashers or tab locks. To avoid assembly errors, reference both rod cap and rod lay as detailed in the "Connecting Rod" section of this chapter.

Once the rod nuts are removed, turn the crankshaft a few degrees to disengage the cap and shank. Using a wooden dowel, drive the piston assembly out the top of the barrel (Fig. 6-34). In event of extreme wear, it might be necessary to ream the cylinder ridge. This operation is described in the following section.

Engines with demountable cylinder barrels trade off easy piston access for difficult connector rod access (Fig. 6-35). With the barrel still assembled, scrap the carbon from the piston top. Bring the piston down to bdc, remove barrel hold-downs, and lift the barrel off the crankcase. A few raps with a rubber mallet might be required

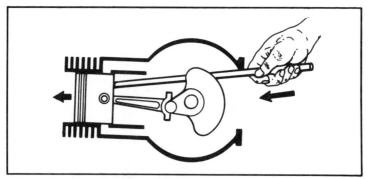

Fig. 6-34. Once the rod cap is detached, tap the piston out of the bore with a wooden dowel placed against the underside of the piston crown. Do not apply force to the upper rod bearing half.

to break the barrel-to-block gasket seal. Raise the piston and stuff the area between the block and rod with clean rags to keep carbon fragments out of the crankcase.

Inspection

Bright rings, uniformly polished and with no vestige remaining

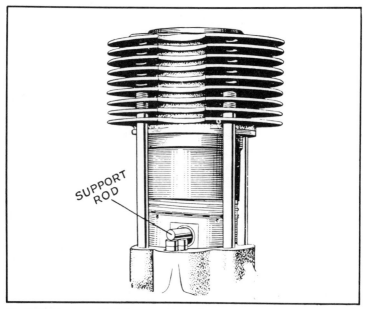

Fig. 6-35. Demountable barrels are lifted off their pistons. They should be supported by a rod or holding fixture.

of tool marks, are simply worn out. Stuck rings, frozen into their grooves, indicate poor maintenance, extreme service, excessive combustion temperature, or consequent loss of ring spring tension. Broken rings have several causes that include inept installation, detonation impact, and worn grooves that allow the rings to twist during stroke reversals.

Chronic detonation can also affect the piston, nibbling at the crown as if mice were at work. The damage usually starts at one edge of the crown—adjacent to the area in the chamber where the fuel charge is slowest to ignite—and progresses toward the center. Check for poor fuel antiknock quality, lean carburetion, excessive ignition advance, and any other condition that would lead to elevated combustion temperature. Excessive load brought on too early in the rpm curve can be a factor because large throttle openings at low speed reduce turbulence and slow flame propagation.

Pre-ignition is rare but obvious when seen from the vantage point of the piston. The center of the crown overheats and can dent or burn through from the combination of high temperatures and premature gas expansion. Check the combustion chamber for any abnormality—such as a hang-nail spark plug thread, a piece of partially detached carbon or a knife-edged exhaust valve that could produce a constant source of ignition. Grinding two-cycle ports oversized for better flow and increased power sometimes has the same results because the bridge between exhaust ports is narrowed and can become incandescent.

Examine the piston skirt for wear. Typically, rubbing contact occurs at two points at right angles to the wrist pin centerline and gradually expands to the whole length of the skirt. Figure 6-36 illustrates abnormal wear patterns produced by bent and twisted connecting rods. Forces that rocked the piston to make these patterns can also drive the wrist pin past its locks and into contact with the cylinder wall.

Deep scratches can mean cylinder bore problems and might smear into the ring grooves, freezing the rings. Light abrasions, giving the piston a matt finish, point to air filter problems. Once sand has been ingested, all bearings become contaminated and the engine should either be scrapped or rebuilt.

Utility and industrial engines are set up fairly tight with piston-to-bore clearances between 0.0015 and 0.002 inches. How much wear is tolerable is, in part, a subjective judgment involving tradeoff between immediate cost and anticipated life to the next overhaul.

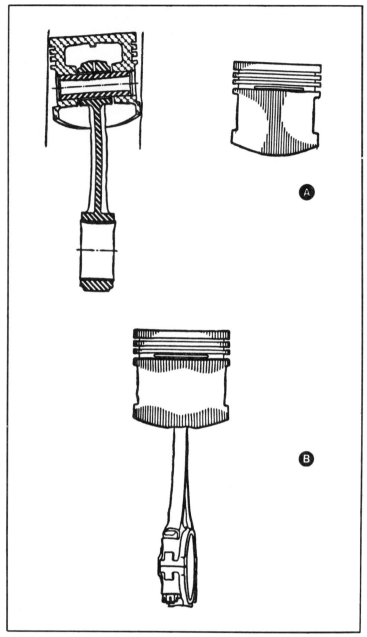

Fig. 6-36. Abnormal thrust face wear, a bent rod tilts the piston in the bore—concentrating wear at the piston ends (A). A twisted rod oscillates the piston producing a wavelike signature (B).

Most factories put the wear limit at 0.005 or 0.006 inches, but small-bore, high-rpm engines are happier if piston clearance does not exceed 0.004 inch.

Pistons usually taper toward their crowns at a rate of about 0.00125 of an inch per inch of height. This allows the hottest part of the piston room to expand. In addition, four-cycle pistons are cam ground so that thrust faces are on the long axis. The piston remains centered on the bore when cold and gradually expands to a full circle as the engine warms to operating temperature. Two-cycle pistons are sometimes round, rather than oval, to control crankcase leakage during startup.

All measurements are made, across the thrust faces, at right angles to the wrist pin. Manufacturers specify a distance above the base of the skirt and just under the wrist pin. Clinton backstops skirt diameter with a measurement across the second ring land (Fig. 6-37).

The final piston check is to determine ring groove width. Remove the rings and scrap all traces of carbon from the grooves, opening oil drain holes in the lowest groove. You might want to use a special groove cleaning tool, available at auto parts stores, or a broken piston ring mounted in a file handle.

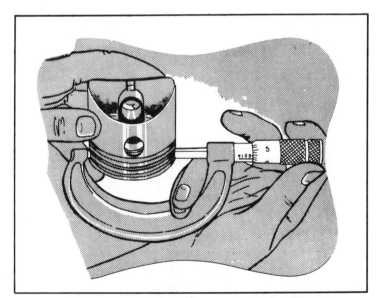

Fig. 6-37. Primary piston measurement is between thrust faces at some specified distance below the wrist pin. Clinton two-cycle straight-cut pistons can also be checked at top ring land.

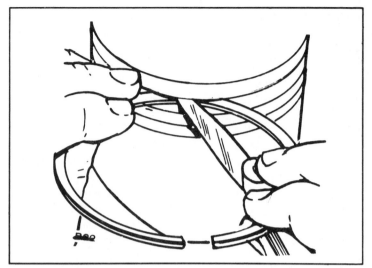

Fig. 6-38. Determine ring side clearance with a new ring as reference. Wear tends to concentrate on the upper side of No. 1 groove. (Courtesy Onan.)

Warning: Piston rings—especially used rings—are razor sharp.

Using a *new* ring, measure side clearance on both compression ring grooves (Fig. 6-38). Excessive side clearance, as defined by the manufacturer, allows the ring to twist during stroke reversals (Fig. 6-39). This condition defects ring sealing geometry and

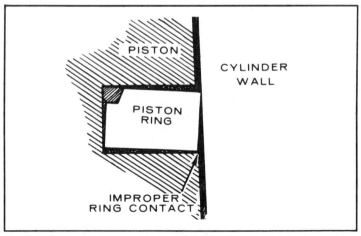

Fig. 6-39. Excessive groove width rounds the leading edge of the ring and can contribute to ring breakage. (Courtesy Onan.)

eventually causes breakage. While it is theoretically possible to recut the grooves overly wide and restore clearance with spacers, the best option is to replace the piston.

Piston Pin

Four-cycle wrist pin bearing wear is almost unheard of because thrust reverses every second revolution. In contrast, two cycle pins are subject to an almost constant downward force that tends to squeeze out what lubrication is present. In either case, the small engine bearing is considered acceptable if it has no perceptible hand-up-and-down plan and if the piston pivots of its own weight.

Pistons incorporate a small offset relative to their pins and some two-cycle piston crowns are shaped to deflect the incoming fuel charge away from the exhaust ports. Consequently, you must install the piston exactly as found. Some are stamped with an arrow or with the letter F (signifying the front of the engine). Others can be oriented by the manufacturer's logo.

Remove and discard the circlips. New circlips are inexpensive insurance against the pin moving into contact with the cylinder bore. If the piston is out of the engine, support it on a wood V-block and drive or press the pin clear of the rod. Do not gouge the pin bore during this process. When the connecting rod remains attached to the crankshaft, it will be necessary to extract the pin with the tool (Fig. 6-40) or by carefully heating the piston. Do not heat with an open flame. Besides inviting a crankcase explosion, this approach is almost guaranteed to distort the piston. Instead, wrap the piston with a rag soaked in hot oil or, less messily, heat the crown with an electric hot plate.

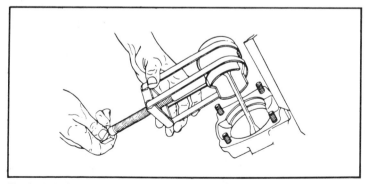

Fig. 6-40. A piston pin extractor is a useful tool that can be purchased through motorcycle or snowmobile dealers. Kohler type is shown.

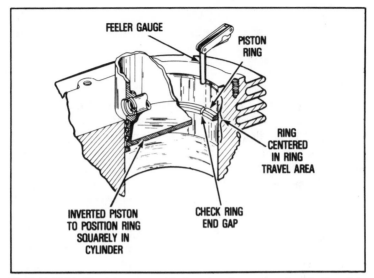

Fig 6-41. Using the piston as a guide, insert each new ring about halfway into the cylinder bore and measure end clearance.

Installation is essentially the reverse process, except that piston pin and pin bores must be well lubricated. Make certain that new circlips seat in their grooves.

Piston Rings

Four-cycle pistons employ three distinct rings. Counting from the bottom, there is the oil control ring (cast in one piece or made up of several steel segments), the scraper, and the top compression ring. Four-ring pistons employ a second, backup compression ring. Replacement ring sets may differ from production sets and, when identical, some manufacturers offer the option of *engineered* replacements. These sets include expanders behind the compression, the scraper, and sometimes the oil control ring to increase ring tension. This expedient permits better conformity with worn bores, but costs something on the order of 2 percent in loss of power and additional fuel consumption.

Two-cycle engines are fitted with two identical compression rings that do, however, have definite upper and lower sides. Installing the rings upside down will cost compression and power. Pegs can be used to secure ring ends to prevent rotation and possible handup in cylinder ports.

Determine the end gap of each ring as verification that the

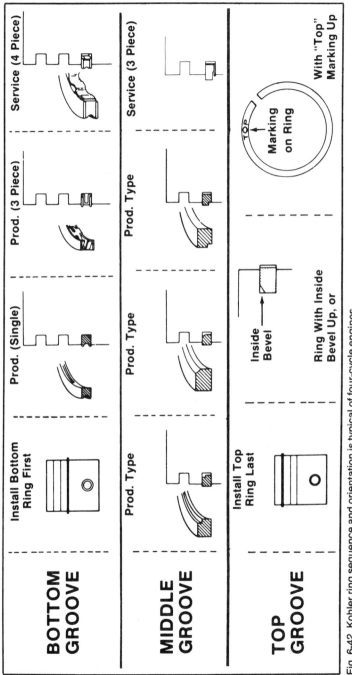

Fig. 6-42. Kohler ring sequence and orientation is typical of four-cycle engines.

correct diameter rings are installed and as a final check on cylinder-bore dimensions. Using the piston crown as a pilot to hold the ring square, insert the ring about midway into the cylinder (Fig. 6-41). Measure end gap with a feeler gauge. Specification varies, but most manufacturers call for about 0.0015 of an inch of ring gap per inch of cylinder diameter. Too large a gap wastes compression and will indicate an undersized ring or oversized cylinder. Too narrow a gap may allow ring ends to abut under thermal expansion, resulting in rapid cylinder wear and premature ring failure. Correct by filing the ends; keep them flat and square.

Lay out rings in the order of installation. Make certain that you have identified each ring and each ring's upper side, which can be marked as such (Fig. 6-42). Using the proper tool, install the oil ring from the top of the piston, spreading the ring only wide enough to clear piston diameter. Repeat this operation for the remaining rings (Fig. 6-43). Verify that rings ride in their grooves and, where applicable, that ring ends straddle locating pegs.

Rotate floating rings to stagger ring gaps some 120° apart so that there will be no clear channel for compression leakage. On Tecumseh engines with relieved valves (shades of Ford hot rod

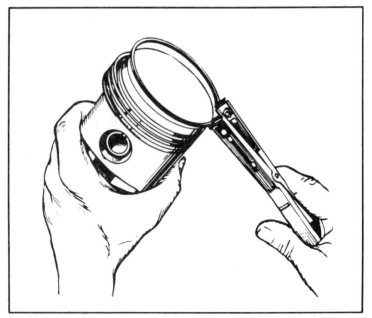

Fig. 6-43. Installing a compression ring on an Onan piston with aid of ring expander.

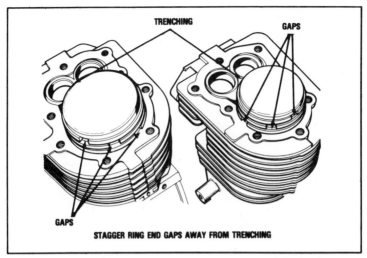

TRENCHING   GAPS

GAPS

STAGGER RING END GAPS AWAY FROM TRENCHING

Fig. 6-44. Rings for Tecumseh engines with relieved, or trenched, valves must be installed with their ends away from the bore undercut.

days), it is important to position ring ends away from the bore undercut (Fig. 6-44).

Installation

Integral Barrel. Turn the crankshaft to the bottom dead center position and press short pieces of fuel hose over rod studs. Lubricate cylinder bore, crankpin, rod bearing, pin and piston (flooding the rings with motor oil). Without upsetting ring gap stagger, install a compressor tool, of the type shown in Fig. 6-45, over the piston. Tighten the band only enough to squeeze rings flush with piston diameter.

Position the piston, attached upper rod as originally found in the engine, and carefully tap the piston out of the compressor. Do not force the issue. If the piston binds, a ring has escaped the tool or there is interference between the rod shank and the crankshaft. Read "Connecting Rod" section that follows before installing the rod cap.

Detachable Barrel. Lubricate cylinder bore, piston pin, and ring areas. Support the piston on the crankcase flange with a rod (as shown back in Fig. 6-35), or with a wooden fork (Fig. 6-46). Some barrels are beveled and can be slipped over the rings without much difficulty. Others are straight cut, and require use of a clamp-type tool shown in Fig. 6-46.

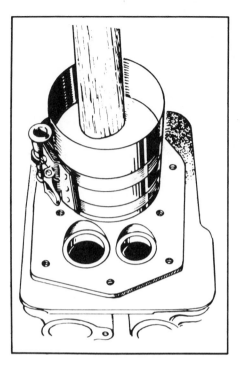

Fig. 6-45. A ring compressor sized for small engines is used when the piston is installed from the top of the bore.

CYLINDER BORES

Pistons for cast-iron engines run directly on block metal without the intermediary of a cylinder liner. Iron provides a fairly durable wearing surface and flows readily enough to mend small scratches. Aluminum-block engines are protected from ring scuff with a layer of chrome applied to the base metal or, as is more often the case, with an iron liner. A few linered bores are chromed for extreme wear and corrosion resistance.

Fig. 6-46. A homemade clamp is easily removed after the barrel is installed over the piston. Inset shows how rings are pegged on some two-cycle engines.

Inspect for deep scratches, aluminum splatter from piston melt, and for chrome separation. The plating is most vulnerable at the top of the bore and around the exhaust ports on two-cycle engines where thermal expansion is greatest. Rechroming the bore is impractical and any evidence of peel means that the block should be scrapped.

Maximum wear occurs near the upper limit of ring travel where heat is greatest, lubrication is minimal, and corrosives are most concentrated. On unchromed bores, wear results in a ridge at the upper limit of ring travel. The amount of ridge is a rough guide to bore wear and can be significant enough to hinder piston extraction. In any event, the ridge must be removed before new rings are installed.

Figure 6-47 illustrates a ridge reamer in use. Adjust cutter tension with the upper nut and rotate the tool clockwise. Insufficient tension dulls the cutter, while too much tension produces chatter and may fracture the carbide cutting edge. Lubricate the tool frequently and stop when the ridge is partly obliterated. No used cylinder is a perfect circle and some evidence of ridge will remain at the long axis.

Cast-iron bores develop a polished glaze that must be removed for new rings to seat (Fig. 6-48). A spring-loaded home that

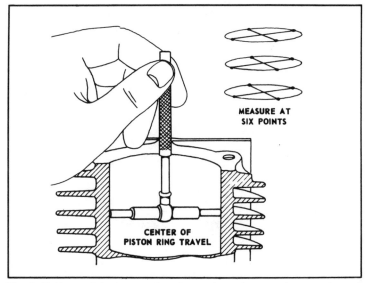

Fig. 6-47. Measure the bore at six points to determine oversize, out of round and taper.

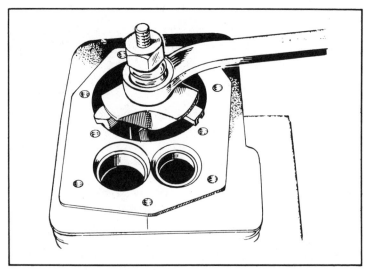

Fig. 6-48. Iron bores develop a ridge at the upper limit of ring travel that must be removed whenever new rings are fitted. (Courtesy Onan.)

automatically conforms to cylinder diameter is preferred for "glaze busting," although patience and sandpaper will do much.

1. Chuck up the hone in a drill press or a 1/2-inch portable drill motor. A smaller drill will provide sufficient power, but most turn at excessively high speeds. Anything more than 400 rpm will produce the "threaded" surface (Fig. 6-49) and defeat the whole operation.

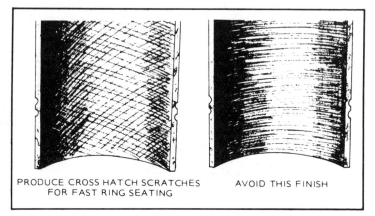

PRODUCE CROSS HATCH SCRATCHES FOR FAST RING SEATING

AVOID THIS FINISH

Fig. 6-49. Honed cylinder should consist of thousands of diamond-shaped points that retain oil and wear-in quickly. (Courtesy Onan.)

2. Fit the tool with a 280-grit stone and lubricate as the manufacturer suggests.

3. Cycle the hone about 70 times a minute with 3/4 inch or so of stone protruding from each end of the cylinder at extremes of travel (Fig. 6-50).

4. Stop when the cylinder bore is uniformly scored.

5. Scrub bore with brush and detergent to remove every trace of abrasive. Cleanup cannot be accomplished with solvent.

Cylinder oversizing is a more serious matter and requires a precision hone. If a lathe is used, the last few thousandths of the cut must be honed to remove tool marks. Most mechanics find it easier to use a hone for the whole operation:

1. Adjust drill press for 300 to 400 rpm spindle speed.

2. Select a coarse (80-grip) stone for ignition cuts.

3. Position the work piece on the tool table (Fig. 6-51). The bore must be vertical and, at the same time, free to move laterally. Figure 6-51 shows an arrangement using shims for vertical alignment. The table can be oiled to further aid centering.

4. Set press stops to extend stones 3/4 of an inch beyond both ends of the bore.

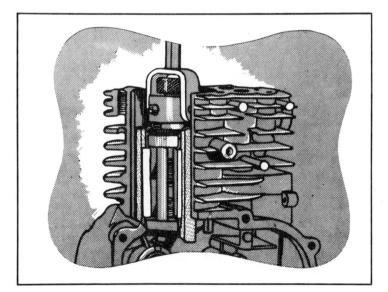

Fig. 6-50. Cutaway view of Clinton block illustrates the cylinder hone at lower travel limit. Note the stone extension beyond the bore.

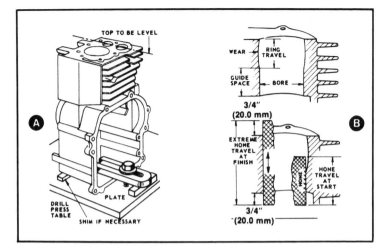

Fig. 6-51. Mount the block or barrel loosely on the drill press table to aid alignment. Vertical-shaft blocks must also be shimmed to bring fire deck level with table (A). When resizing, begin at the lower and least worn portion of the bore (B).

5. Adjust the hone to conform with the lower bore diameter (which will be smaller than upper bore). Contact should be positive, but not so firm as to prevent turning the tool by hand.

6. Lubricate stones as the manufacturer suggests. Petroleum solvents can dissolve the binder and this causes rapid stone wear.

7. Star press and keep hone moving about 70 strokes a minute. As lower cylinder enlarges, adjust hone for greater reach. Eventually the whole length of the cylinder will be traversed on each stroke.

8. Keep a close eye on how much metal is removed. The bore might tend to bell mouth at the ends.

9. Clean felts and replenish lubricant at frequent intervals.

10. When the bore is straight and within 0.002 inch of final size, stop and change to a 280 finishing stone.

11. Make the final cut; stop frequently for a dimension check. Verify running clearance against replacement piston. The actual diameter may vary 0.0005 inch for nominal diameter.

12. Scrub the bore with a brush, hot water, and detergent. Wipe dry with paper shop towels. Scrub until the towels are no longer stained with abrasive. Oil immediately.

CONNECTING RODS

Aluminum is the material of choice for four-cycle connecting rods that, almost always, are split at the big end. Utility and light industrial do not have replaceable bearings; crank and piston pins run against the rod itself. Figure 6-52 illustrates the standard pattern that, in this case, incorporates an oil slinger below the cap for splash lubrication.

Better-quality engines might use precision bearing inserts at the big end in conjunction with a bushing at the small end (Fig. 6-53). Undersized inserts (0.010 and 0.020 inch for American-made engines) allow the crank to be reground.

Traditionally, two-cycle connector rods were steel forgings with needle bearings at both ends. However, light and moderate-output plants, including some outboard motors, currently use aluminum, (which can be fitted with races for needle bearings or which can

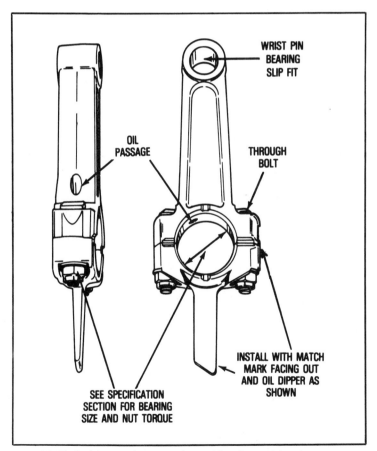

Fig. 6-52. Typical four-cycle connecting rod has integral bearings.

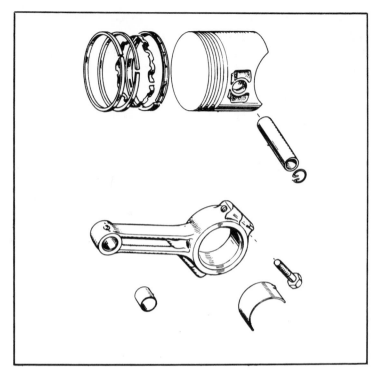

Fig. 6-53. Onan conn rod features a bushed small end and precision inserts at the big end.

itself act as the bearing). Plain bearing engines are adequate for light duty in applications where exhaust smoke is of little consequence. These engines require as much as 1 part of oil to 24 parts of fuel. As a point of comparison, an automobile engine is considered worn out of oil consumption is 1 to 400.

Figure 6-54 illustrates a connector rod for a two-cycle industrial power plant. The aluminum rod rides on a bushing at its upper end and on single-row or double-row needles at the crank end. Note the use of replaceable races.

Catastrophic rod failure almost always originates at the big end. How this happens is, in part, a function of big end bearing type. Plain bearings skate on a pressurized wedge of oil that appears soon after startup. Once up to speed, the bearing should, in a sense, hydroplane and make no direct contact with its journal.

Insufficient clearance between the crankpin and bearing prevents the oil wedge from forming; excessive clearance allows the wedge to leak faster than it can be formed. In either case, the result is metal-to-metal contact, fusion, and a broken connecting rod.

Needle bearings make rolling contact against their races without the cushion of an oil wedge. Consequently, any imperfection—fatigue flaking, rust pitting, "skid marks"—means bearing seizure and rod failure.

Split big ends occasionally crumple into bit-sized chunks as a result of insufficient rod-bolt torque. Proper torque might not have been applied during assembly or rod locks might have given way, allowing the bolts to shake loose. This is why manufacturer's torque specifications must be followed to the letter and why new lock washers or locknuts must be installed whenever the rod is disassembled. Bend-over tab locks usually carry a spare tab that can be employed during the first overhaul. Once the tab is engaged, it cannot, in conscience, be straightened and reused.

Orientation

Correct orientation is vital and, counting the piston, has three aspects:

● Piston to rod. The piston pin may be offset relative to the bore and two-cycle pistons may be asymmetrical.

● Rod assembly to engine. Some connector rods are drilled for oil and vapor transfer; others are configured so that reverse installation locks the crankshaft.

● Cap to rod. In order to maintain the necessary precision, most engine makers assemble the rod and cap and ream to size.

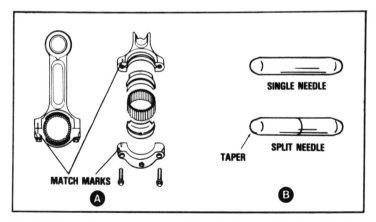

Fig. 6-54. Two-cycle rod employs needle-bearing lower end for durability. Note that split needles assemble with squared ends together.

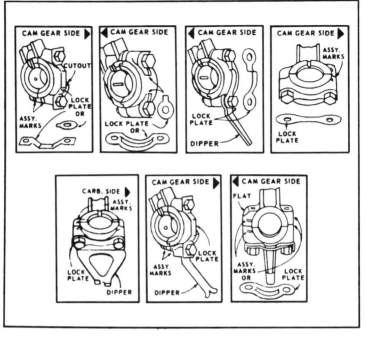

Fig. 6-55. Briggs & Stratton rod-to-engine and cap-to-rod orientation. Another example of embossed rod and cap index marks can be seen in A of Fig. 6-54. McCulloch engines employ a steel rod that is fractured after machining; when cap is installed correctly, the break becomes almost invisible.

Stamped or embossed marks identify cap orientation (Fig. 6-55). Failure to assemble the cap correctly results in early and catastrophic failure.

Inspection

Modern pistons are "free floating." The piston assembly should pivot on the rod at room temperature. Piston pin-to-rod clearance should be in the neighborhood of 0.0002 of an inch with more than 0.0005 or 0.0006 as the wear limit on plain bearings. Needles can tolerate a little more play. Because needle bearings are part of the two-stroke repertoire, excessive clearance can mean trouble. Two-cycle engines get very little lubrication in this area and, what lubrication there is, is poorly distributed. These engines run under more or less constant compression. This condition tends to starve the lower part of the bearing. Note that small end needles will be located by steel thrust washers.

The big end gets most attention. Inspect the bearing for blue temper marks—always a sign of serious trouble—and for surface flaws. Small scratches can be tolerable in a low-rpm, plain-bearing application. Any imperfection—including tiny pits, rust or needle imprints on antifriction crankpins, races or rollers—is grounds for immediate rejection. Replace all needles as a set. A new needle in an old set will ride higher than the others and take the full brunt of the load.

Needle bearings are "measured" by assembling the rod dry with match marks aligned and by determining play by feel, and bearing condition by sound, or more exactly, absence of scrapes, rattles and other sounds of protest. Plain bearings must be actually measured to determine running clearance, out of round, and taper.

The traditional way of doing this is to measure the crankpin diameter across bottom dead center and at 90° from bdc at each end of the pin. This shows out of round because the pin might be egg-shaped, and taper, since one end of the pin might have a larger average diameter than the other. In general, 0.001 of an inch of out of round and taper are about the maximum allowed. Some mechanics are more concerned about taper than out of round. Once the crankpin diameter is known, the rod is assembled, match marks—as always—together, and its diameter is measured across bdc and normal to it.

The difference between average crankpin and rod bearing diameter equals running clearance, which is subject to specification, but which should fall between 0.001 of an inch or a new and somewhat tight assembly to, say, about 0.004 (which is pushing things a bit). Do not attempt to restore bearing clearance by filing the ends of the rod cap. The expedient does not work for very long.

Another way to establish bearing clearance and crankpin asymmetry is to use plastic-gauge wire (available from auto parts jobbers). The soft, plastic wire is precisely dimensioned and flattens as the bearing cap is installed. Wire width converts to bearing clearance via a scale on the package. Follow this procedure:

1. Turn the crankshaft and assembled rod to bottom dead center.

2. Remove the rod cap.

3. Wipe off all oil on rod cap and exposed crankpin.

4. Tear off a piece of gauge wire and lay it along the full length of the crankpin (A of Fig. 6-56).

5. Install the rod cap, oriented correctly, and pulled down evenly to factory torque specifications.

Caution: Do not rotate the crankshaft during this procedure.

6. Remove the cap and measure the width of the gauge wire against the scale printed on the envelope (B of Fig. 6-56). Average width corresponds to bearing clearance; variations in width from one end of the crankpin to the other show taper.

7. Repeat the process, using two pieces of gauge wire, as shown in C of Fig. 6-56. This is a cross-check on taper and indicates out of round.

Even if you prefer to make the initial determination with precision gauges that seem to show taper and out of round more positively than wire, the final check on the installed bearing clearance should be made with plastic. This is a positive measurement with very little room for errors of interpretation.

ASSEMBLY

Coat upper and lower bearing surfaces thoroughly and liberally with clean motor oil. Failure to do this can ruin a bearing on initial startup. The insert type of big-end bearings sometimes have an oil hole that defines the upper end. Otherwise bearing inserts interchange between rodcap and shank. Uncaged needle bearings can be fixed around the periphery of the crankpin with grease or, following the old practice, with beeswax. Protect the crankpin, during piston installation, with short lengths of fuel line over the rod bolts.

Check the piston-to-block, piston-to-rod, and rod-cap orientation one final time. Turn the crank down to bdc and, using your fingers, guide the rod assembly home. Install the correctly orientated cap, new rod locks, and run the bolts down to specified torque, keep the cap square during the process.

Pull the engine over by hand for several revolutions to detect possible binds. The rod should move easily from side to side along the crankpin. Most manufacturers do not provide a side play specification, but the rod is comfortable with several thousandths of an inch of axial freedom.

CRANKSHAFTS

It is always good practice to align timing marks before four-cycle engines are disassembled. Crankshaft and camshaft timing marks index at top dead center on the compression stroke.

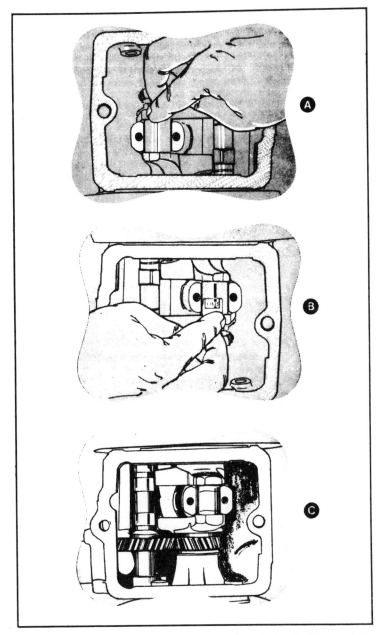

Fig. 6-56. First, lay a piece of plastic gauge wire along the length of the crankpin (A), install the cap to factor torque, remove the cap, and measure wire gauge width against scale on envelope (B). These repeat the operation using two pieces of gauge wire around the fore and aft perimeter of the crankpin (C).

Secondary marks on rotating balance or accessory-drive shafts are indexed to the crank or cam after primary alignment is made.

Occasionally, timing marks wear away and the mechanic must time the engine from the "rock" position. Rotate the crankshaft to bring No. 1 piston to top dead center on what will become the compression stroke. Install the camshaft; it should slip easily under the tappets. Rock the crankshaft a degree or two on each side of tdc, alternately engaging the intake and exhaust valves. Timing is correct when freeplay splits evenly between the two valves. If one valve leads the other, reposition the camshaft one tooth from that valve.

Crankshafts that run on plain, main bearings extract easily without interference with the camshaft. Timing marks should be clearly visible from the top side of the block (A of Fig. 6-57). Some engines

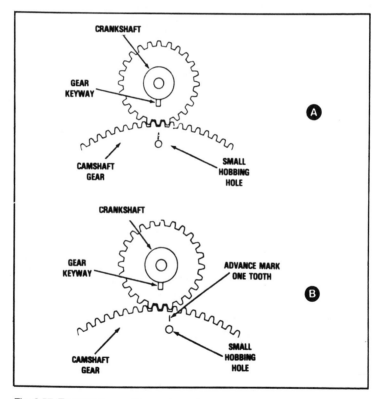

Fig. 6-57. Tecumseh uses the crankgear key as a timing referent. On all models except one, this indexes with a stamped mark on the cam gear (A). The exception is the Craftsman-label engines with an elemental carburetor that are advanced one tooth (B).

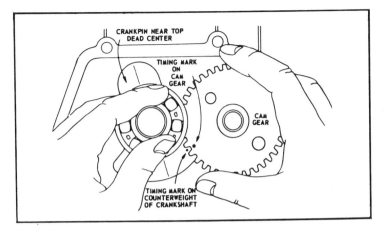

Fig. 6-58. Axle-supported camshaft on engines with antifriction main bearings must be dropped out of mesh for crankshaft access. Briggs & Stratton is shown and Clinton is similar.

employ the crankshaft gear keyway as one of the marks. Tecumseh-made Craftsman engines with fixed-adjustment carburetors are advanced one tooth for sake of mid-range torque (B of Fig. 6-57). As far as I know, they are the only small engine deliberately mistimed.

Antifriction (ball or tapered roller) bearing cranks can present something of an extraction problem. The top-side bearing rides in a carrier and, because of limited space, the camshaft must be dropped out of position to maneuver the crankshaft throw out of the block. Timing marks on the crankshaft side often take the form of a chamfered tooth or may be stamped on the counterweight (Fig. 6-58).

Displace the camshaft by driving out the cam axle through the magneto side of the block (Fig. 6-59). Note the expansion plug that should be oil-proofed with sealant before assembly. Timing might be easier if the associated crankshaft gear tooth is marked with chalk or a crayon. This is particularly true on Briggs & Stratton 30400 and 320400 models.

Inspection. Figure 6-60 illustrates inspection points for a Briggs & Stratton plain bearing crankshaft. Other makes do not have the integral point cam, represented by the flat on the magneto end, and there is normally no need to measure journal-bearing diameter when antifriction bearings are fitted. These journals do not wear unless the race has spun (in which case the crank may not be salvageable). The crankshaft shown has an integral gear, that is usually a separate part on other makers. Crank and cam drive gears should be renewed as pairs.

Pay special attention to the crankpin. Check for out-of-round, taper, and concentric wear as described under "Connecting Rods." Remember that deep scratches are grounds for rejection. As mentioned in that section, needle bearing crankpins require a glass-smooth surface without the slightest hint of galling or rust pitting. When oversized connecting rod bearing inserts are available, the crankpin can be ground to fit. Otherwise repair is by way of replacement.

Check oil ports for burrs and shavings that could restrict oil flow to the bearings. Crank journals should be lightly polished before assembly with 600 wet-or-dry emory cloth saturated in oil. The best way to do this is to cut a strip of sand cloth to journal width, wrap it around the journal, and spin with a shoe lace of leather thong. Remove all traces of abrasive from the crank and, particularly, the oil drillings.

No small engine manufacturer allows crankshafts to be straightened. This is primarily because of the legal ramifications of a broken crankshaft and attached lawnmower blade or whatever.

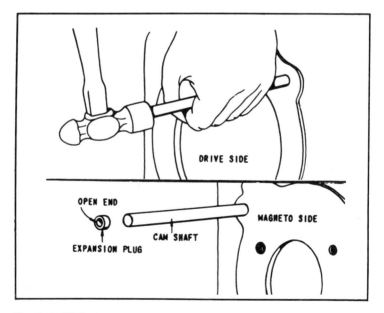

Fig. 6-59. While the crankshaft-side timing mark is usually on the gear, exigencies of ball-bearing mains sometimes make this impossible. Then the timing mark is on crankshaft counterweight.

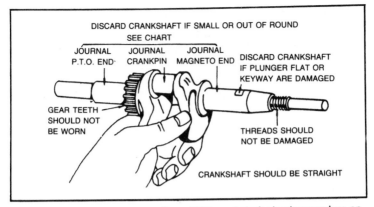

Fig. 6-60. With only a few exceptions, B&S crankshaft check procedure applies to other makes as well.

Another consideration is that a crankshaft bend—normally confined to the top-side stub and extends out of the block—can displace the counterweights, bowing the crankpin in the process.

Experienced and patient craftsmen can straighten cranks bent a few thousandths. If you want to pursue the matter, recognize, dear reader, that you are on your own and that describing this process is not an invitation to perform it.

The work requires two machinist's vee-blocks, two dial indicators, and a straightening fixture that is usually built around a hydraulic jack. The crank is supported by the blocks at the main bearings and the indicators are positioned near opposite ends of the shaft. Total runout should be no more than 0.001 of an inch (or 0.002 of an inch indicated). Using the fixture, the crank is brought into tolerance in small increments with frequent checks. Once the indicators agree, the crank is then sent out for magnetic particle inspection to detect possible cracks. Skipping this final step, which only costs a few dollars at an automotive machine shop, can be disastrous for all concerned.

Upon assembly, check the crankshaft end play. Depending upon block construction, this check is made internally (Fig. 6-61) or with a dial indicator from outside of the engine. The amount of float is not crucial so long as parts have ample expansion room. Typical specs fall in the 0.003-inch-to-0.005-inch range. Several engine makers supply thicker base or bearing cover gaskets for use when the float has been absorbed by a new crank or flange. A thrust washer—usually placed between crank and topside main and, occasionally, on the magneto side—compensates for wear.

CAMSHAFTS

The camshaft rides on an axle pin (as shown back in Fig. 6-58) or else is supported by plain bearings at the magneto side of the block and the top-side cover. The latter type can be removed and installed without compressing valve springs if first turned to the timing mark index position.

Most camshaft failure is obvious: Once the surface hardness goes, the lobes wear, the round, gear teeth break, or the gear fragments. A careful mechanic will measure valve lift and bearing clearance. This is particularly true if the cam extends through the engine to serve as a power takeoff. In most cases, cam drive side bearings are placeable when wear exceeds 0.005 of an inch or so.

The cam might include a compression release to aid starting. Briggs & Stratton displaces the cam laterally when the starter engages to unseat the exhaust valve. Little can go wrong with this device and no special service procedures are required. Tecumseh and Kohler employ centrifugal compression releases that prevents exhaust valve closure below 500 or 600 rpm (Fig. 6-62). Check the

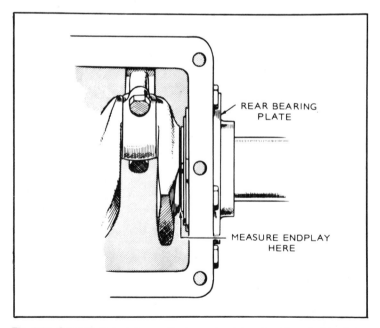

REAR BEARING PLATE

MEASURE ENDPLAY HERE

Fig. 6-61. Crankshaft end play, or float, may be determined from inside of the engine with feeler gauge between crank cheek and thrust bearing (as shown on this Onan engine). It is more convenient to check flange-type engines externally, at the pto stub.

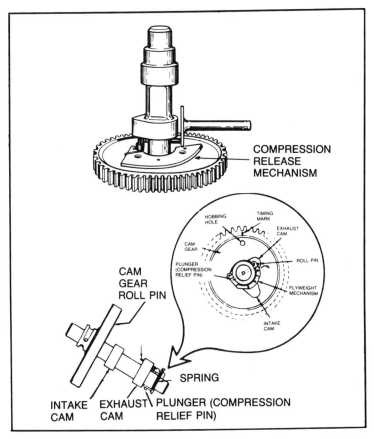

COMPRESSION
RELEASE
MECHANISM

HOBBING
HOLE

TIMING
MARK

EXHAUST
CAM

CAM
GEAR

ROLL PIN

PLUNGER
(COMPRESSION
RELIEF PIN)

FLYWEIGHT
MECHANISM

INTAKE
CAM

CAM
GEAR
ROLL PIN

SPRING

INTAKE
CAM

EXHAUST
CAM

PLUNGER (COMPRESSION
RELIEF PIN)

Fig. 6-62. Centrifugal compression release fitted to some Tecumseh engines should be checked during overhaul and, if necessary, replaced as a unit with the camshaft.

operation by hand and look for wear on pivots and weight stops. Clinton and other manufacturers sometimes employ a cam-actuated advance mechanism on engines with side-mounted magnetos. Verify operation by hand. Remove springs only as necessary for replacement.

MAIN BEARINGS

The crankshaft runs against plain or antifriction bearings or a combination of both types with an antifriction bearing at the top end. Plain bearings can be made of brass. In such cases they are relatively easy to replace or are integral with an aluminum block. Antifriction bearings usually are present as ball or roller bearings with inner cones and outer races (cups) to protect both the crankshaft and the castings. Some two-cycle engines have used needle bearings (riding directly on the crank). Antifriction bearings should be replaced at first sign of roughness and as part of every engine rebuild.

Antifriction. Figure 6-63 illustrates the more or less typical setup using two tapered roller bearings with a washer and shims at the top side to control end play. Check by removing all traces of lubricant from the bearings and spinning the outer races by hand. Roughness or the tumbrel-like noise of loose cones means that the bearing should be renewed.

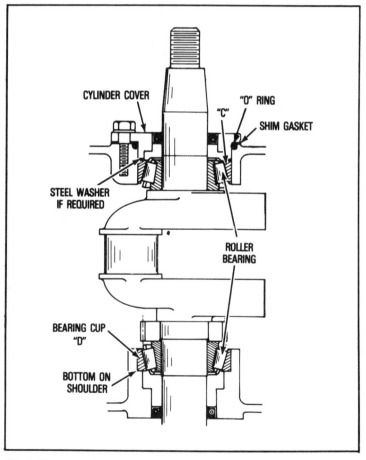

Fig. 6-63. Tecumseh roller bearing arrangement is typical of small industrial engine practice. Shim gasket and optional washer determine crankshaft float.

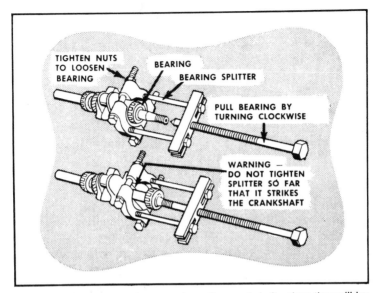

Fig. 6-64. Antifriction bearings remain on the crankshaft unless they will be replaced. (Courtesy Clinton.)

Caution: Do not spin antifriction bearings with compressed air. In addition to damage from water in the air source, the turbine effect will overspeed the bearings.

Remove the bearing from the crankshaft with aid of a bearing splitter (Fig. 6-64). Once drawn in this manner, bearings cannot be reused. The preferred method of installation is to heat the bearing in a container of oil until the oil begins to smoke (a condition that corresponds to a temperature of about 375° F).

The bearing should be supported away from the bottom of the container with a wire mesh. The more usual method is to press the bearing cold by supporting the crankshaft at the web and applying force to the inner race only. Figure 6-65 illustrates this operation for Kohler double-press fit bearings. First the bearing is pressed into its cover with the arbor against the outboard race, and then the cover assembly is installed with press force confined to the inner race. Antifriction bearings seat flush against the shoulders provided. Check end play against specification and adjust as necessary with gaskets or shims.

Antifriction bearings are hardware items that can be purchased from bearing-supply houses at some savings over dealer prices. However, be certain that the replacement matches the original in all respects. Unless you have certain information to the contrary,

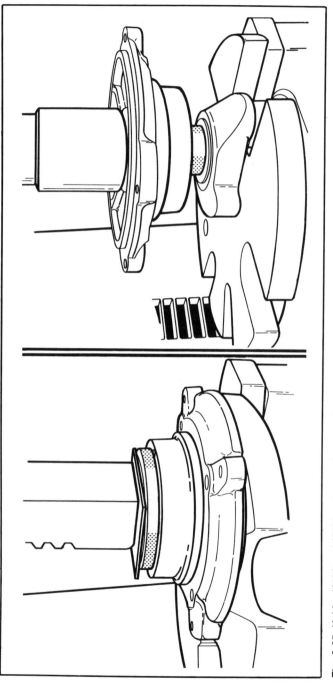

Fig. 6-65. Kohler K482 and K532 series engines employ pto bearings with double interference fit. The cup is pressed into bearing cover, and then the inner race is pressed on the crankshaft. Supporting crankshaft web protects the crankpin.

do not specify the standard C1 clearance for bearings with inner races. Ask for C3 or C4; they are looser fits to allow room for thermal expansion.

Plain. Determine bearing clearance with inside and outside micrometers and compare with factory specs for the engine in question. Most are set up with 0.0015 inch new clearance and tolerate some 0.0035 inch before rework.

All engines from major manufacturers can be rebushed, but this is not a do-it-yourself project. The work is best left to a dealer who has access to the necessary reamers, pilots, and drivers.

Thrust bearings are normally present as a hardened washer at the top end of the crankshaft. Kohler and other manufacturers sometimes specify a proper babbit-coated or roller thrust bearing. Poorly maintained vertical-shaft engines will develop severe galling

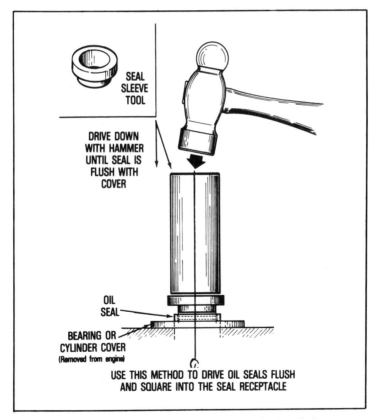

Fig. 6-66. Always use the correct sized driver to ensure that installation stresses are confined to the outer edge of seal retainer.

at the flange thrust face that can be corrected by resurfacing the flange or replacing the casting.

SEALS

Seals, mounted outboard of the main bearings, contain the oil supply for four-cycle engines and seal crankcase pressure in two-strokes. Seal failure can be recognized by oil leaks at the crankshaft exit points or, on two-cycle engines, by hard starting and chronically lean fuel mixtures. Seals should be replaced during an overhaul and must be replaced to protect the investment of a rebuilt engine.

Install replacement seals with the maker's mark visible and the steep sides of the lip toward the pressure. Lubricate the lip with light grease and, unless the seal is already covered with an elastomer, coat the metal rim OD with gasket sealant. Be careful not to allow the sealant to spread to the lips or the oil return port just inboard of the seal.

Installation is best done with a factory seal driver that concentrates force on the rim OD. A length of pipe of the appropriate dimension will suffice. Drive the seal to the original depth (usually flush or just under flush), unless the crankshaft seal area is worn. In that case, adjust seal depth to engage an unworn area on the crank, but do not block the oil return port in the process.

The crankshaft must be taped during installation to protect seal lips from burrs, keyway edges, and threads. Celophane tape, because it is relatively thin, works best.

GOVERNOR MECHANISMS

The unit shown in Fig. 6-67 is typical of the crankcase part of most mechanical governor assemblies. Paired flyweights, driven at some multiple of engine speed by the camshaft, pivot outward with increasing force as rpm increases. This motion is translated into vertical movement at the spool and appears as a restoring force on the carburetor throttle linkage. Work the mechanism by hand, checking for ease of operation and obvious wear. The governor shaft presses into the block or flange casting and, in event of replacement, must be secured with Loctite bearing mount and installed at the proscribed height.

OILING SYSTEMS

Two-cycle engines are, of course, lubricated by oil mixed with

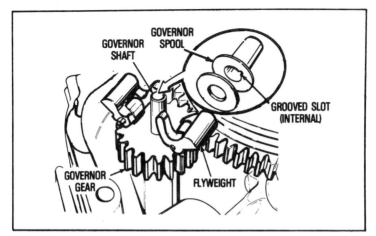

Fig. 6-67. Typical centrifugal governor flyweight assembly employs plastic spool.

the fuel. Bearings may be vented with ports or milled slots to encourage oil migration.

Four-cycle systems can be more complex and worthy of special attention. Indeed, some of the best mechanics seem to spend an inordinate amount of time tracing the circuits, cleaning those that are accessible with rifle bore brushes and compressed air, and making certain that blanking plugs are secure and pressure-tight.

Any of the three oiling systems are used. Most small, side-valve engines depend upon a *splash system*, in which crankcase oil is agitated by a dipper mounted on the connector rod cap or camshaft gear. Briggs & Stratton engines in this class employ a camshaft-driven slinger (Fig. 6-68). Other than thoroughly cleaning the inside of the crankcase, checking dipper orientation (another reason to make certain the rod cap is on right) and inspecting the slinger for wear, no special maintenance procedures are required.

Because splash systems develop no pressure, lubrication depends upon proximity and gravity return to the crankcase. Drilled passages are not present to deliver oil to remote areas and neither an oil filter or gauge can be fitted.

Semi-pressure systems combine splash with positive feed to some bearings. The Tecumseh system, used on vertical crankshaft engines, is fairly typical of the breed (Fig. 6-69). A small plunger-type pump (Fig. 6-70), driven by the camshaft, draws oil from a port on the cam during the pump intake stroke. As the plunger telescopes closed, a second port on the camshaft hub aligns with the pump barrel and oil is forced through the hollow camshaft to

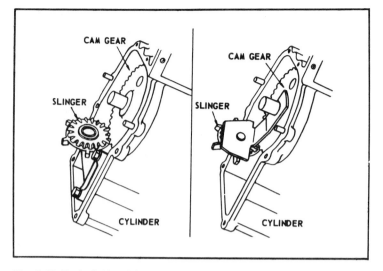

Fig. 6-68. Early (left) and late-production Briggs & Stratton oil slingers. The later variant may incorporate centrifugal governor weights and a wave washer between bracket and flange. Bracket should be replaced with camshaft hole ID is 0.49 of an inch or larger.

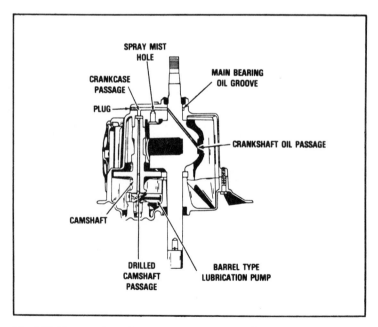

Fig. 6-69. Tecumseh semipressure system pumps oil to the upper main bearing and (optionally) to the crankpin.

254

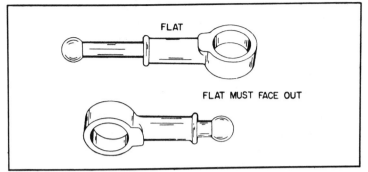

Fig. 6-70. The plunger pump shown in extended (suction) and collapsed (discharge) positions. Note that the flat faces out on assembly.

a passage on the magneto side of the block. Cross drillings in the camshaft provide lubrication to the bearings. Once in the block passage, the oil flows around a pressure-relief valve (set to pen at 7 psi) and into the upper main bearing well. Most models feature

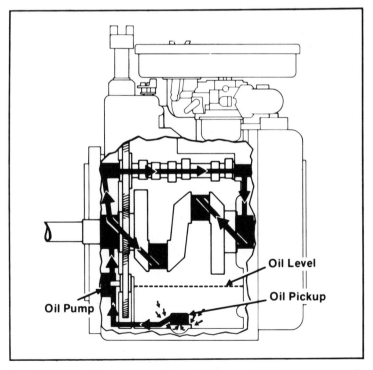

Fig. 6-71. Kohler full-pressure system splits the crankshaft into separate oiling circuits.

a crankshaft drilling to provide oil to the crankpin.

Blow out the passages with air and check the pump for scores and obvious wear. Replace pump plunger and barrel as a matched assembly.

Caution: The pump must be assembled with the flat side out and primed with clean motor oil before startup.

Some Tecumseh engines use an Eaton-type oil pump, recognized by its star-shaped impellor. Check for scuffing on the impellor and pump case ID. Clearance between impellor and pump cover should be gasketed to 0.006-0.007 of an inch. Except for relocation of the pressure relief valve in the flange between the pump and camshaft, oil circuitry is as previously discussed.

Full-pressure systems deliver pressurized oil to all crucial bearing surfaces, although some parts receive lubrication from oil thrown off the crankpin (cylinder bore, cam gear) or by oil flowing back to the sump (valve guides). The Kohler system, used on KT17 Series II and KT19 Series II engines, is typical of most, (although circuitry varies between engine makes and models). A conventional gear-type pump supplies oil to the top side main, No. 1 crankpin and to the camshaft that serves as a gallery to bring oil to the magneto-side main bearing and No. 2 crankpin. A pressure-relief valve, under the top-side main bearing carrier, limits pressure to 50 psi to prevent bearing erosion. Provision for an oil pressure sendor is by way of a 1/16-inch NPTF plug on the top-side of the crankcase. See Fig. 6-71.

Index

Index

A

adjustment tabs, 88
adjustments, high-speed mixture, 86
adjustments, idle rpm, 87
adjustments, low-speed mixture, 87
alignment, simple, 23
alternator, 179
angleich, 106
armature, 168, 171
automatic advance mechanism, 28

B

barrel, detachable, 229
barrel, integral, 229
batteries, 155
battery capacity, 36
battery polarity, 180
battery, 36, 178
battery, lead-acid, 157
battery, nicad, 159-63
bearing, thrust, 170
bearings, integral, 235
bearings, main, 3, 247
bearings, thrust, 251
Bendix carburetor cleaner, 95
Bendix, 165-67
bore, 10, 231
bores, iron, 232
Bosch magneto, 57
Bosch, 165, 170, 172
brake horsepower, 13
breather, 215

Briggs & Stratton, 33, 40-42, 45, 49,
    51-54, 60, 62, 70, 76, 96, 99,
    110-13, 118, 119-20, 126,
    140-43, 130, 145, 157, 160-64,
    167, 173-74, 187, 193-96, 204,
    207-08, 211, 238, 243, 254
bushings, 169
bushings, throttle shaft, 96
bushings, throttle, 100

C

camshaft, 4, 112, 219, 246
carbon deposits, 92, 193
carburetor cleaner, 95
carburetor elements, 67
carburetor nomenclature, 68
carburetor, downdraft, 76
carburetor, float, 73, 78
carburetor, lean-running, 94
carburetor, rich-running, 94
carburetor, suction lift (fuel pump),
    70
carburetor, suction-lift, 69
carburetor, updraft, 76
carburetors and fuel systems, 67
carburetors, cleaning and repair of,
    95
carburetors, float, 80
carburetors, needle and seat
    assemblies for, 96
carburetors, side-draft, 77
carburetors, temperamental, 85

carburetors, throttle slide, 91
carburetors, troubleshooting, 91
carburetors, types of, 69
castings, 101
charging systems, 178
choke, automatic, 183
Chrysler engines, 139
Chrysler West Bend engine, 84
circlips, 225
circuit, high-speed, 67
circuit, low-speed, 67
circuits, starting, 155
clamp, homemade, 230
Clinton, 5, 56, 70-71, 78, 179, 210,
    217, 223, 233, 243, 249
clutch assemblies, 123
clutch shoes, Fairbanks-Morse, 123
clutch, 169
coil air gap, 64
coil and battery ignition, 183
coil and battery, 34
coils, electromagnetic field, 164
coils, em, 165
commutator, 170
compression ratio, 11-12
compression, 5
connecting rods, 234
contact points, 43
crankcase, 7, 8, 18, 115, 218
crankpin, 241
crankshaft shoulder, 45
crankshaft threads, 43
crankshaft, 3, 5, 17, 25, 43-45, 219,
    239-250
crankshaft, horizontal, 218
crankshaft, vertical, 218
cylinder block, 3
cylinder bores, 230
cylinder head, 1, 192-94
cylinder, 10
cylinder, honed, 232

D

Delco-Remy, 176
diaphragm pumps, 113
diodes, 174, 179

E

Eaton rewind starter, 119
Eaton, 120, 127-28
electrical system, 155-56
engine idle, 90
engine knocks, 188
engine maintenance, 1
engine mechanics, 185
engine overhaul, 191

engine power output, 189
engine timing, 24
engine, Chrysler West Bend, 84
engine, diagnosis, 186
engine, loop-scavenged, 9
engine, Magna-Matic, 57
engine, rebuilding an, 191
engine, single-cylinder, 2
engine, two-cycle, 7
engines, air-cooled, 20
engines, basics of, 1
engines, Fairbanks-Morris, 134
engines, four-cycle, 6
engines, motorcycle, 10
engines, operation of, 4
engines, single-cylinder Kohler, 16
engines, single-cylinder utility, 69
engines, two-cycle, 226
engines, two-stroke-cycle, 8

F

Fairbanks-Morse, 134-39
fields, 172
filter, air, 21
filter, Kohler, 105
filters, air, 102
filters, oil-bath, 103
filters, polyurethane, 104
flange, 4
float bowl, 102
flywheel gear, 169, 175
flywheel hub, cracked, 44
flywheel keys, damaged, 44
flywheel knocker, 43
flywheel puller, 42
flywheel, 40-41, 47, 55-56, 61-62,
    118, 146, 163, 167-68, 178,
    188, 194
fuel intake, 5
fuel pump, 72, 108, 118
fuel starvation, 93
fuel systems and carburetors, 67

G

gases, exhaust, 6
gases, expansion of, 6
gear lash, 147
governor arm, 109
governor mechanisms, 252
governors, 104
governors, air vane, 104
governors, centrifugal, 105-07
governors, fixed-speed, 106

H

Harley-Davidson, 79
horsepower, 14

hp, indicated, 13
hp, taxable, 13
hydrometer, using a, 158

I

idle, refusal to, 93
ignition coils, 55
ignition system output, insufficient, 33
ignition system, Magna-Matic, 49
ignition system, troubleshooting the, 35
ignition systems, conventional, 34
ignition, 23, 187
integral barrel, 229

J

Jacobsen, 23-24, 117

K

Kohler engine, 75
Kohler, 25, 29, 35, 75, 92, 95, 105, 113, 165, 183, 187, 194, 198, 202, 205, 214, 218, 227, 249-56

L

lean roll, 90
lubrication, 17
lubrication, splash, 19

M

magneto ignition, 182
magneto stator, 24
magneto systems, 37
magneto, 27, 47, 55-56, 61
magneto, Bosch, 57
magneto, parts of a, 37-38
magneto, troubleshooting a, 38
Magneton, 64
Magnetron CDI, retrofit of a, 60
Magnetron, 63
mainspring, 132, 140, 143, 150-51
motor generator, 176
motor-generator, polarized, 177
motor-generators, wiring, 176

O

ohmmeter, checking diodes with an, 174
oil consumption, 190
oil filters, 103
oil pump, 4, 255-56
oil slingers, 254
oil, 244
oil, motor, 240
oil, splash system, 253

oil, weight requirements of, 18
oiling systems, 252
Onan charging system, 181
Onan, 12, 15, 19, 26, 28, 36, 68, 73, 112-14, 156, 171, 195, 224, 228, 236, 246

P

performance data, 13
Permatex, 64
Phelon, 27
piston orientation, 237
piston pin, 3, 225
piston position, 32
piston rings, 1, 224, 226
piston rods, 222
piston, 1, 5, 12, 220
pistons and rings, 215
pistons, inspection of, 238
plugs, expansion, 100
plugs, welch, 100-01
plunger wear, excessive, 53
point alignment, 50
point contacts, 52
point disassembly, 52
point gap adjustment, 25, 51
point-to-cam adjustment, 26
points and condenser troubleshooting, 48
points, 39
points, filing, 55
points, gap, 54
puller, hub, 41
pump, fuel, 108, 110
pump, primer, 100
pumps, diaphragm, 113
pumps, mechanical, 110

R

rectifier, 172
relay and solenoid, 175
ring, compression, 228
rings and pistons, 215
rings, 221
rod, connecting, 3-5
rods, 220
rods, connecting, 234

S

seals, 252
sheave, pivot shaft, 131
sheave, split, 131
sheaver, one-piece, 133
solenoid and relay, 175
solenoid, 156, 183

solid-state systems, 57
spark plug gap, 39
spring installation, 125
spring, balance, 109
spring, rewind, 124
springs, valve, 197-98
starter motors, 163, 175
starter, Briggs & Stratton side-pull, 123, 198
starter, Eaton heavy-duty, 130
starter, Eaton rewind , 129
starter, Eaton, 132
starter, Fairbanks-Morse, 135
starter, gear-driven, 145
starter, horizontal-engagement, 145
starter, rewind, 122
starter, troubleshooting, 118
starter, utility, 137
starter, vertical-pull, vertical-engagement, 148-49
starters, rewind, 117
starters, service procedures for re-wind, 119
starters, side-pull, 117
starters, vertical pull, 137
stator pedestal, 24
stroke, compression, 6
stroke, exhaust, 6
stroke, power, 6

T

Tecumseh CDI system, the, 59
Tecumseh overhead valve gear, 197
Tecumseh, 10, 24, 41, 61, 81, 110, 144-46, 149, 162, 166, 173, 180-82, 192-96, 242, 247-48, 254-56
throttle opening, 92
throttle, 67

Tillotson, 80-89
timing drill details, 31
timing drill, Kohler single-cylinder, 29
timing light, 27
timing mark, 244
timing methods, 23, 32
timing procedures, 24
torque, 13-14

V

valve gear modification, 214
valve gear, 197
valve guides, 206, 209
valve inspection, 200
valve lash adjustment, 213
valve lash, 214
valve lathe, 203
valve nomenclature, 199
valve seal, 206
valve seats, 210-12
valve seats, loose, 211
valve, scrap, 212
valves, 4, 194
Viton seat, 97
voltage regulator, 181
voltage, cranking, 159

W

Walbro, 96-98
West Bend, 24
Wico, 24, 27, 56
Wisconsin Robin, 25

X

Xenon lamp, 34

Z

Zenith small-engine carburetors, 75

Edited by Steven Bolt

Other Bestsellers From TAB

☐ **66 FAMILY HANDYMAN®
WOOD PROJECTS**

Here are 66 practical, imaginative, and decorative projects ...literally something for every home and every woodworking skill level from novice to advanced cabinetmaker: room dividers, a freestanding corner bench, china/book cabinet, coffee table, desk and storage units, a built-in sewing center, even your own Shaker furniture reproductions! 210 pp., 306 illus. 7″ × 10″.
**Paper $14.95   Hard $21.95
Book No. 2632**

☐ **DESIGNING, BUILDING AND
TESTING YOUR OWN SPEAKER
SYSTEMS . . . WITH
PROJECTS — 2nd
Edition—Weems**

You can have a stereo or hi-fi speaker system that rivals the most expensive units on today's market ... *at a fraction of the ready-made cost!* Everything you need to get started is right here in this completely revised sourcebook that includes everything you need to know about designing, building, *and* testing every aspect of a first-class speaker system. 192 pp., 152 illus.
Paper $10.95   Book No. 1964

☐ **BUILD A PERSONAL EARTH
STATION FOR WORLDWIDE SAT-
ELLITE TV RECEPTION—2nd
Edition—Traister**

You'll find a thorough explanation of how satellite TV operates ... take a look at the latest innovations in equipment for individual home reception (including manufacturer and supplier information) . . . get detailed step-by-step instructions on selecting, assembling, and installing your earth station . . . even find a current listing of operating TV satellites (including their orbital positions and programming options).
**Paper $14.95   Hard $21.95
Book No. 1909**

☐ **THE DARKROOM BUILDER'S
HANDBOOK—Hausman & DiRado**

Have your own fully-equipped home darkroom for a fraction of the price you'd expect to pay! Picking and choosing from the features incorporated into the book's darkroom examples, you'll be able to design a "dream" darkroom that meets your own specific needs. Among the highlights: choosing a suitable location, the equipment and accessories you'll need, even how you can make your home darkroom pay for itself! 192 pp., 238 illus. 7″ × 10″.
**Paper $12.95   Hard $21.95
Book No. 1995**

☐ **ADVANCED AIRBRUSHING
TECHNIQUES MADE SIMPLE—
Caiati**

Here are all the professional tips and tricks needed to achieve the full spectrum of airbrushing effects—for retouching black-and-white and color photos, art and illustration work, mixed media, mural production, modeling and dilorama finishing, vignetting and photo montage procedures, and more! Highlighted by more than 165 illustrations. 144 pp., 168 illus. plus 27 color plates. 7″ × 10″.
Paper $19.95   Book No. 1955

☐ **PROFESSIONAL PLUMBING
TECHNIQUES—ILLUSTRATED
AND SIMPLIFIED—Smith**

This plumber's companion includes literally everything about plumbing you'll ever need! From changing a washer to installing new fixtures, it covers installing water heaters, water softeners, dishwashers, gas stoves, gas dryers, grease traps, clean outs, and more. Includes piping diagrams, tables, charts, and arranged alphabetically. 294 pp., 222 illus.
Hard $16.95   Book No. 1763

Other Bestsellers From TAB